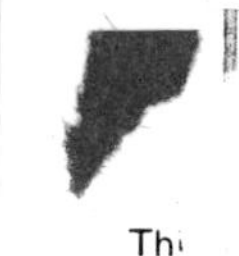

ANDRE

Th

LONGMAN BOTANY HANDBOOK

the elements of plant science illustrated and defined

GMAN YORK PRESS

The author

ANDREW SUGDEN studied botany at Oxford University. He devoted the subsequent decade to botanical research in South American tropical forests. Since 1986 he has been the Editor of the monthly review journal *Trends in Ecology & Evolution*.

Acknowledgement

The author wishes to thank the staff and students (1980-1) of the Botany School of the University of Oxford for their helpful comments and advice on the definitions in this book.

YORK PRESS
Immeuble Esseily, Place Riad Solh, Beirut.

LONGMAN GROUP UK LIMITED
Longman House, Burnt Mill, Harlow,
Essex CM20 2JE
and Associated Companies throughout the world

First published 1984 as *The Longman Illustrated Dictionary of Botany*
This edition first published 1992
ISBN 0 582 09965 X

Illustrations by Charlotte Styles
Photocomposed in Britain by Prima Graphics, Camberley
Produced by Longman Group (FE) Ltd.
Printed in Hong Kong

Contents

How to use the handbook

This handbook contains over 1200 words used in the botanical sciences. These are arranged in groups under the main headings listed on pp. 3–4. The entries are grouped according to the meaning of the words to help the reader to obtain a broad understanding of the subject.

At the top of each page the subject is shown in bold type and the part of the subject in lighter type. For example, on pp. 18 and 19:

18 · **CELLS**/MEMBRANES, ORGANELLES

CELLS/MEMBRANES, ORGANELLES · **19**

In the definitions the words used have been limited so far as possible to about 1500 words in common use. These words are those listed in the 'defining vocabulary' in the *New Method English Dictionary* (fifth edition) by M. West and J. G. Endicott (Longman 1976). Words closely related to these words are also used: for example, *characteristic*, defined under *character* in West's *Dictionary*.

1. To find the meaning of a word

Look for the word in the alphabetical index at the end of the book, then turn to the page number listed.

The description of the word may contain some words with arrows in brackets (parentheses) after them. This shows that the words with arrows are defined near by.

(↑) means that the related word appears above or on the facing page;

(↓) means that the related word appears below or on the facing page.

A word with a page number in brackets (parentheses) after it is defined elsewhere in the handbook on the page indicated. Looking up the words referred to in either of these two ways may help in understanding the meaning of the word that is being defined.

The explanation of each word usually depends on knowing the meaning of a word or words above it. For example, on p. 80 the meaning of *peduncle*, *pedicel*, and the words that follow depends on the meaning of the word *inflorescence*, which appears above them. Once the earlier words are understood those that follow become easier to understand. The illustrations have been designed to help the reader understand the definitions but the definitions are not dependent on the illustrations.

2. To find related words

Look in the index for the word you are starting from and turn to the page number shown. Because this handbook is arranged by ideas, related words will be found in a set on that page or one near by. The illustrations will also help to show how words relate to one another.

For example, words relating to cell division are on pp. 45–50. On p. 45 *cell division* is followed by words used to describe mitosis; pp. 46 and 47 give words used in describing chromosomes; pp. 48 and 49 illustrate and explain meiosis; and p. 50 gives words relating to chromosome numbers.

3. As an aid to studying or revising

The handbook can be used for studying or revising a topic. For example, to revise your knowledge of photosynthesis, you would look up *photosynthesis* in the alphabetical index. Turning to the page indicated, p. 32, you would find *photosynthesis*, *autotrophic*, *heterotrophic*, *chloroplast*, and so on; on p. 33 you would find *grana*, *lamellae*, and so on. Turning over to p. 34 you would find *Calvin cycle* etc.

In this way, by starting with one word in a topic you can revise all the words that are important to this topic.

4. To find a word to fit a required meaning

It is almost impossible to find a word to fit a meaning in most dictionaries, but it is easy with this book. For example, if you had forgotten the word for the outer whorl of the perianth of a flower, all you would have to do would be to look up *perianth* in the alphabetical index and turn to the page indicated, p. 70. There you would find the word *calyx* with a diagram to illustrate its meaning.

5. Abbreviations used in the definitions

abbr.	abbreviated as	p.	page
adj	adjective	pl.	plural
e.g.	*exempli gratia* (for example)	pp.	pages
etc.	*et cetera* (and so on)	sing.	singular
i.e.	*id est* (that is to say)	v	verb
n	noun	=	the same as

THE HANDBOOK

phytochemistry (*n*) the chemistry of plants. **phytochemical** (*adj*).

atom (*n*) the smallest unit of a chemical element (↓). Atoms contain electrons (↓), protons (↓) and neutrons (↓). The numbers of electrons and protons in an atom are equal. **atomic** (*adj*).

the four commonest atoms in biological compounds

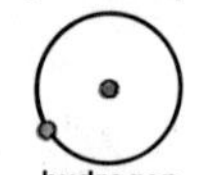

hydrogen (1 proton, 1 electron)

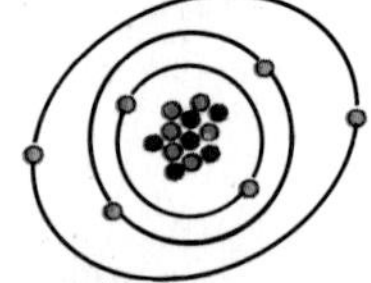

carbon (6 protons, 6 neutrons, 6 electrons)

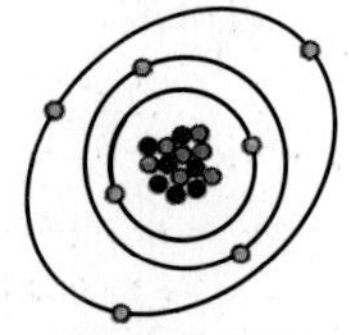

nitrogen (7 protons, 7 neutrons, 7 electrons)

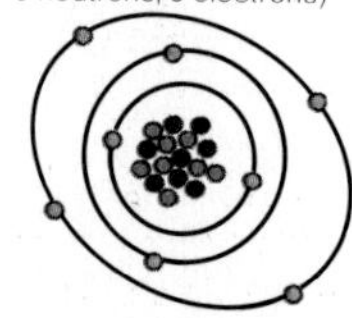

oxygen (8 protons, 8 neutrons, 8 electrons)

element (*n*) a substance consisting of atoms (↑) all of the same kind. An element cannot be changed into another element except by splitting atoms. Each element, e.g. oxygen, carbon, or nitrogen, has its own characteristic number of protons (↓) in its atoms.

proton (*n*) a particle with a positive electric charge, found in all atoms (↑). The electric charge of a proton is exactly equal and opposite to that of an electron (↓), so an atom has no charge. A proton is a hydrogen ion (↓), since hydrogen atoms consist of only one proton and one electron.

electron (*n*) a particle with a negative electric charge, found in all atoms (↑). The electric charge on an electron is exactly equal and opposite to that on a proton (↑). The addition or removal of electrons from atoms creates ions (↓). Electrons are 1840 times lighter than protons.

neutron (*n*) a particle with no electric charge, found in all atoms (↑) except hydrogen atoms. Neutrons have the same weight as protons (↑).

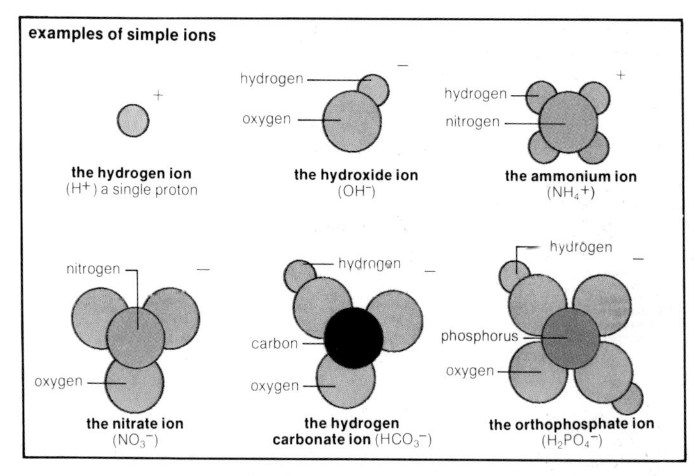

ion (*n*) an atom (↑) or molecule (↓) with an electric charge, due to having an unequal number of protons (↑) and electrons (↑). An ion with a positive charge has more protons than electrons, and an ion with a negative charge has less protons than electrons. **ionization** (*n*).

molecule (*n*) the smallest unit of an element (↑) or compound which occurs naturally. Molecules consist of more than one atom (↑). A molecule of hydrogen consists of two hydrogen atoms (H_2), and a molecule of carbon dioxide has one atom of carbon and two of oxygen (CO_2). **molecular** (*adj*).

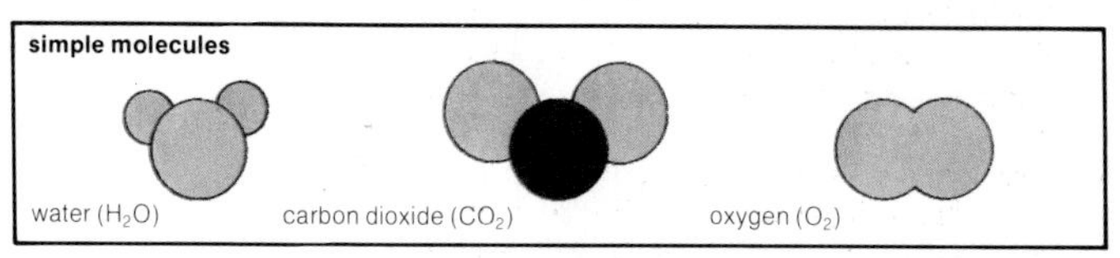

macromolecule (*n*) a large molecule (↑) consisting of many atoms, e.g. proteins (p. 56), nucleic acids (p. 51), polysaccharides (p. 30).

crystal (*n*) a solid symmetrical (p. 71) structure made of molecules (↑) all of the same kind and size. **crystalline** (*adj*).

compound[1] (*n*) a molecule (p. 9) which consists of more than one kind of atom (p. 8).

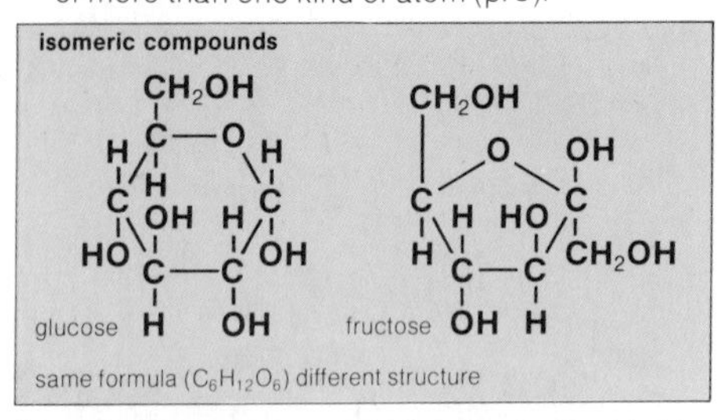

isomers (*n*) two or more molecules (p. 9) with the same numbers and kinds of atoms (p. 8), but with different arrangements of atoms and sometimes different chemical properties. **isomeric** (*adj*).

polymer (*n*) a chemical substance formed by the joining together of many molecules (p. 9) of the same kind, e.g. polysaccharides (p. 30), polypeptides (p. 56) and nucleic acids (p. 51).

monomer (*n*) one of the units in a polymer (↑).

biological polymers—the macromolecules

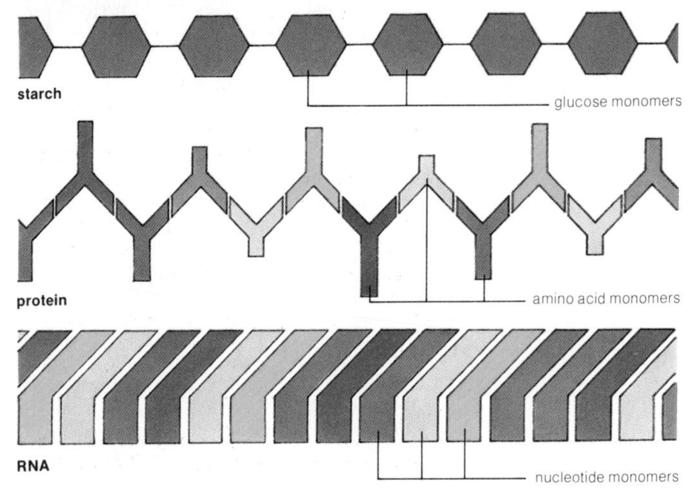

reduction (*n*) the process in which a substance (1) gains electrons (p. 8); (2) has oxygen removed from it; or (3) has hydrogen added to it. **reduce** (*v*), **reductive** (*adj*).

oxidation (*n*) the process in which a substance (1) loses electrons (p. 8); (2) has oxygen added to it; or (3) has hydrogen removed from it. **oxidize** (*v*), **oxidative** (*adj*).

redox (*adj*) of chemical reactions that involve oxidation (↑) and reduction (↑).

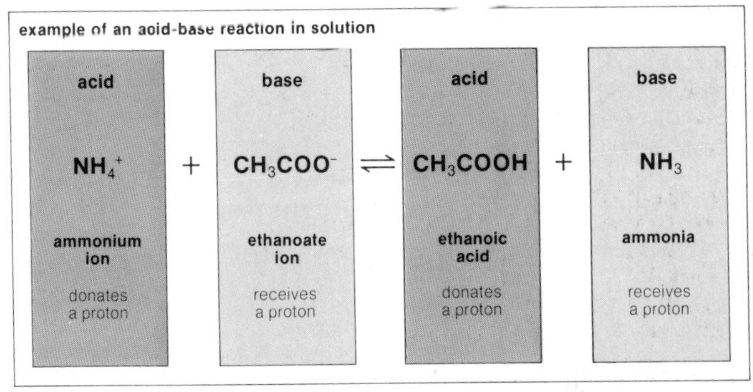

reaction (*n*) chemical process in which two or more compounds act on each other, by exchanging atoms (p. 8) or electrons (p. 8), to produce different compounds.

potential energy energy that is stored in a molecule (p. 9) and can be released to drive chemical reactions. Potential energy is usually measured in terms of electric charge.

organic (*adj*) of compounds which contain atoms (p. 8) of carbon. The compounds synthesized (p. 13) by living organisms (p. 118) are organic.

inorganic (*adj*) of compounds which do not contain carbon.

acid (*n*) any chemical compound which can donate protons (p. 8) to water molecules (p. 9). The acidity of a solution (p. 12) is measured on the pH scale (-log H^+ concentration). **acidic** (*adj*).

base[1] (*n*) any substance which can accept protons (p. 8) from water molecules (p. 9). **basic** (*adj*).

solution (*n*) a liquid with substances dissolved (↓) in it.
dissolve (*v*) *of solids* to break down into molecules (p. 9) or ions (p. 9) when placed in a liquid.
solute (*n*) a substance which is dissolved (↑) in a liquid.
solvent (*n*) a liquid in which substances are dissolved (↑).
soluble (*adj*) of substances which can be dissolved (↑), e.g. sugar in water. **solubility** (*n*).
insoluble (*adj*) not soluble (↑).
aqueous (*adj*) of solutions (↑) in which the solvent (↑) is water.
concentration (*n*) the amount of a substance dissolved (↑) in a given volume of liquid.
evaporation (*n*) the process by which molecules (p. 9) of a liquid become a gas. **evaporate** (*v*).

solution

solute
solvent
solid
liquid
solution
liquid

hydrolysis (*n*) a chemical reaction involving the breakdown of a molecule (p. 9) into two molecules, with the addition of the parts of a molecule of water. **hydrolyze** (*v*), **hydrolytic** (*adj*).

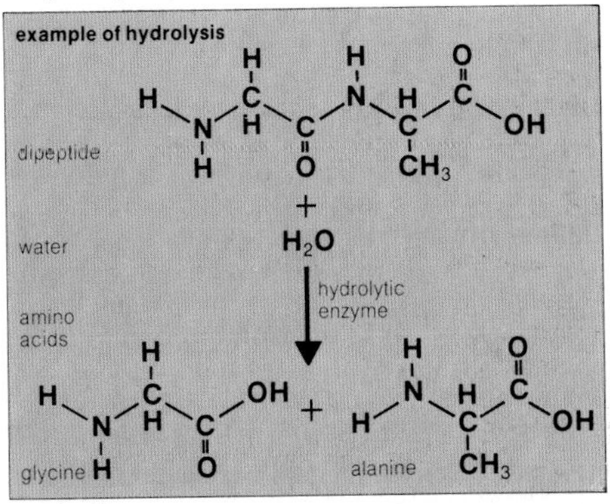

example of hydrolysis

synthesis (*n*) the process of building chemical compounds from small molecules (p. 9), e.g. carbohydrates (p. 28) from carbon dioxide and water in photosynthesis (p. 32), or proteins (p. 56) from amino acids (p. 56) in protein synthesis (p. 57). **synthesize** (*v*), **synthetic** (*adj*).

phosphate (*n*) an inorganic (p. 11) ion (p. 9), present in the soil, which is an important nutrient (p. 111) for plants. Phosphate (PO_4^{3-}), is used in the synthesis (↑) of ATP (p. 26) during photosynthesis (p. 32) and respiration (p. 22). It is also used in the nucleotide (p. 52) molecules (p. 9) of nucleic acids (p. 51).

nitrate (*n*) an inorganic (p. 11) ion (p. 9), present in the soil, which is an important nutrient (p. 111) for plants. Nitrate (NO_3^-) provides nitrogen for the synthesis (↑) of amino acids (p. 56) and other nitrogen-containing compounds, e.g. nucleotides (p. 52).

orthophosphate (*n*) another term for inorganic (p. 11) phosphate (↑) ion (p. 9). Pi (*abbr.*).

ammonia (*n*) an inorganic (p. 11) molecule (p. 9) with one atom (p. 8) of nitrogen and three atoms of hydrogen (NH_3).

metabolism (*n*) the sum of the chemical reactions which occur in an organism or a cell. Metabolism involves the breakdown of organic (p. 11) compounds, releasing energy that is used in the synthesis (p. 13) of other compounds. **metabolize** (*v*).

metabolite (*n*) a substance produced by metabolism (↑).

metabolic pathway a set of chemical reactions which follow each other in a sequence. Each reaction uses the product of the reaction before it. *See also* **metabolism** (↑).

inhibitor (*n*) a substance which prevents or slows down a chemical reaction or process. Some inhibitors can slow down enzyme (↓) reactions by blocking the active site (↓) of the enzyme. **inhibit** (*v*).

inhibition (*n*) the prevention or slowing down of a metabolic (↑) reaction, e.g. by an inhibitor (↑) or by temperatures that are too high or too low.

feedback (*n*) the process by which a product at or near the end of a metabolic pathway (↑) affects the reactions (p. 11) at the beginning of the same pathway. Feedback can be either positive or negative.

inhibition

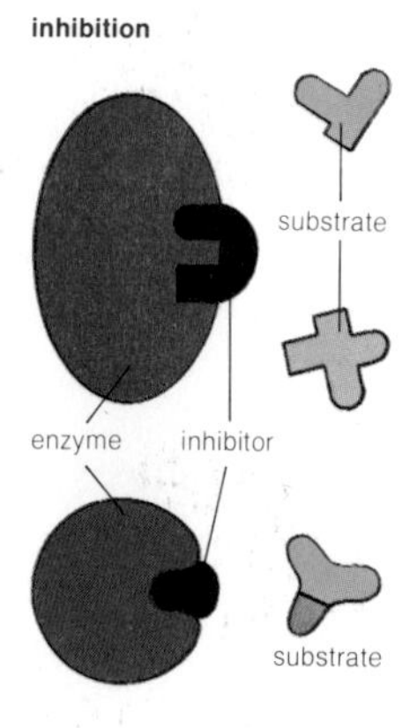

examples of feedback

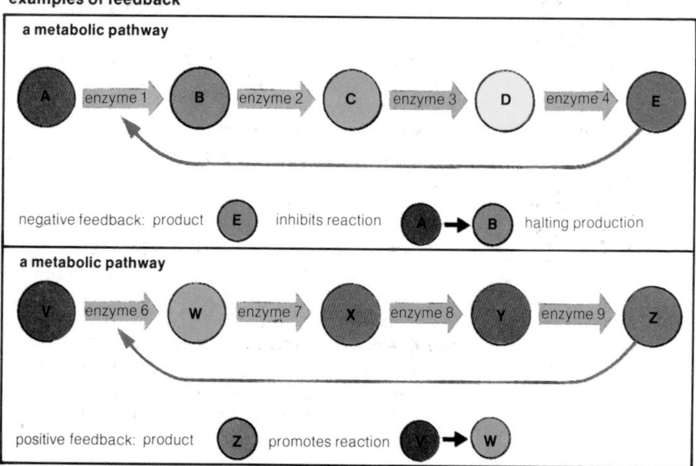

the function of enzymes in catalysis of reactions

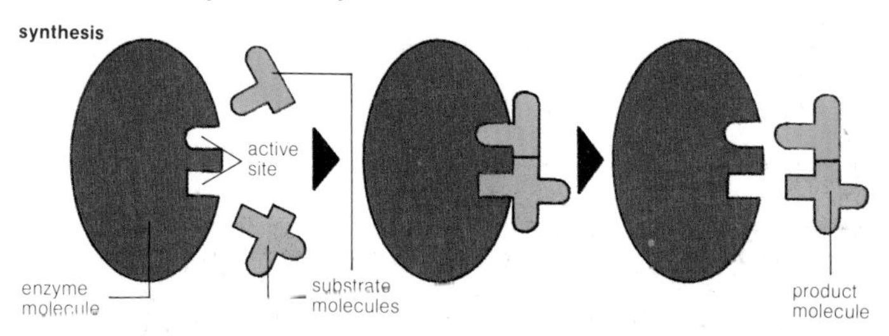

enzyme (*n*) a protein (p. 56) which, in very small quantities, catalyses (↓) and controls the natural chemical reactions of metabolism (↑). Enzymes are usually large complex molecules (p. 9), and most are responsible for one or two particular reactions in the cell. Cells contain many thousands of different enzymes.

substrate[1] (*n*) the general name for the substance on which an enzyme (↑) acts.

catalysis (*n*) the process in which natural chemical reactions are made more rapid, e.g. by enzymes (↑). **catalyze** (*v*), **catalytic** (*adj*).

catalyst (*n*) a substance that increases the rate of a chemical reaction without itself being changed in the process, e.g. enzymes (↑).

active site the part of an enzyme (↑) molecule (p. 9) to which substrate (↑) molecules become attached and where catalysis (↑) takes place.

coenzyme (*n*) a non-protein (p. 56) substance which some enzymes (↑) require to make them active. Different enzymes have different coenzymes, e.g. vitamins (↓).

vitamin (*n*) an organic (p. 11) substance required as a coenzyme (↑) in many of the chemical reactions of metabolism (↑). There are many different kinds of vitamin, and they are required by organisms in very small amounts.

multi-enzyme complex a set of enzymes (↑), often grouped into a particular arrangement in an organelle (p. 16) or a membrane (p. 18), which catalyze (↑) different reactions in the same metabolic pathway (↑).

cell (*n*) a unit of protoplasm (↓) surrounded by a membrane (p. 18). Nearly all living organisms are made of one or more cells. Cells are either prokaryotic (↓) or eukaryotic (↓). Plant cells differ from animal cells by having cell walls (↓), and plastids (p. 18) in the case of eukaryotic cells. **cellular** (*adj*).

eukaryotic cell (plant)

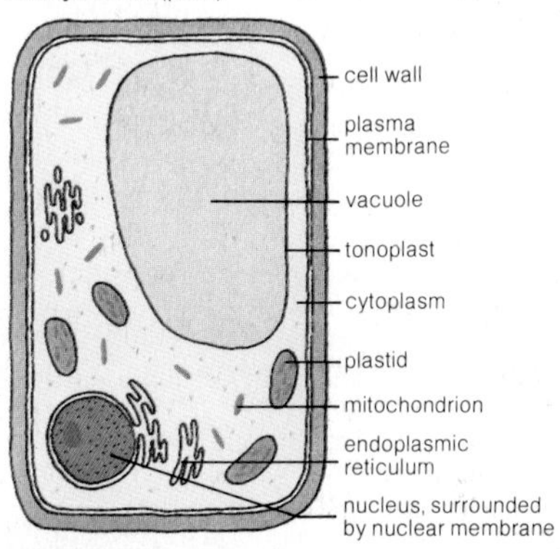

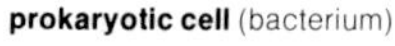

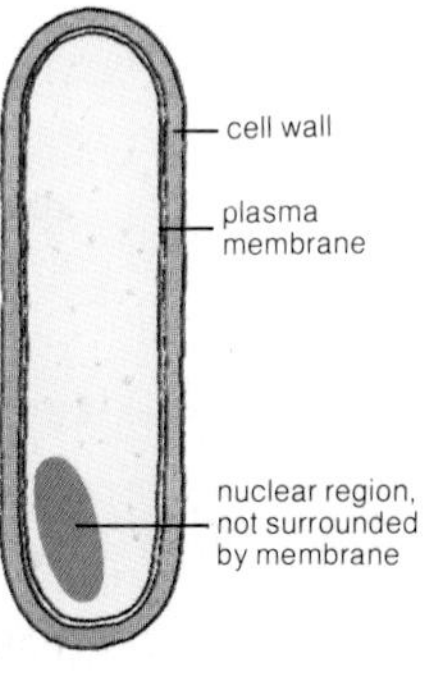

prokaryotic (*adj*) of cells having no organelles (↓) and no membrane (p. 18) surrounding the nuclear (p. 19) material of the cell, e.g. bacteria (p. 119). **prokaryote** (*n*).

eukaryotic (*adj*) of cells with a nucleus (p. 19) surrounded by a nuclear membrane (p. 19), and distinct organelles (↓). **eukaryote** (*n*).

organelle (*n*) a body inside a eukaryotic (↑) cell, usually with a membrane (p. 18) around it. There are usually several kinds of organelle in any cell, and each kind has a special function. For example, the function of chloroplasts (p. 32) is photosynthesis (p. 32), the function of mitochondria (p. 21) is respiration (p. 22).

cytology (*n*) the study of cells by microscopy.

intracellular (*adj*) inside a cell.

extracellular (*adj*) outside a cell.

protoplasm (*n*) the general name for all the substances and bodies inside a cell. All living organisms are made of protoplasm.

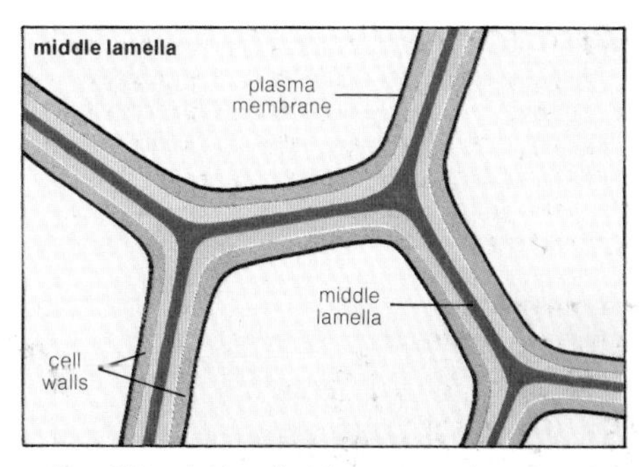

cell wall the rigid wall which surrounds a plant cell, lying outside the cell membrane (p. 18). Cell walls are made mainly of carbohydrate (p. 28) polymers (p. 10) such as cellulose (↓). All plants, fungi (p. 163) and bacteria (p. 119) have cell walls, but animals do not.

microfibril (*n*) one of the threads of carbohydrate (p. 28) polymer (p. 10) of which cell walls (↑) are made.

middle lamella the thin young cell wall (↑) formed between two new eukaryotic (↑) cells after cell division (p. 45). The middle lamella is made of pectin (↓), and the thicker layers of cellulose (↓) are laid down on either side of it.

cellulose (*n*) a carbohydrate (p. 28) polymer (p. 10) made of glucose (p. 28) molecules, which is the most important substance in plant cell walls (↑).

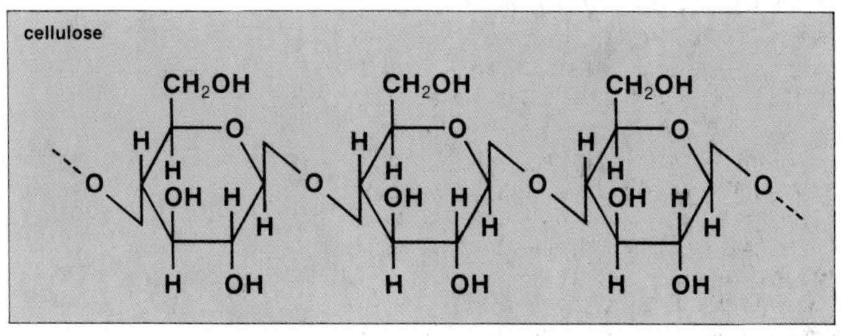

pectin (*n*) an acidic polysaccharide (p. 30) found in young cell walls (↑). **pectic** (*adj*).

membrane (*n*) a thin sheet of soft material which protects and encloses cells and organelles (p. 16). Membranes control the movement of substances in and out of cells and organelles. Biological membranes are made of protein (p. 56) and phospholipid (p. 31).
cell membrane the membrane (↑) which encloses a cell.
plasmalemma (*n*) = the cell membrane (↑).
plasma membrane = the cell membrane (↑).
protoplast (*n*) a plant cell or a bacterial (p. 119) cell, not including the cell wall (p. 17).
cytoplasm (*n*) all parts of a cell outside the nucleus (↓) and inside the cell membrane (↑).

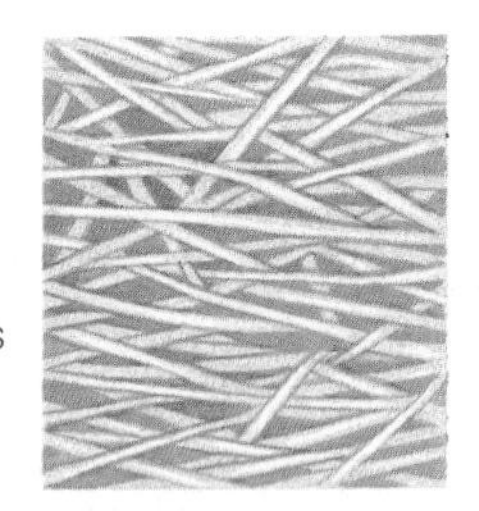
cellulose microfibrils in surface view of plant cell wall (× 24,000)

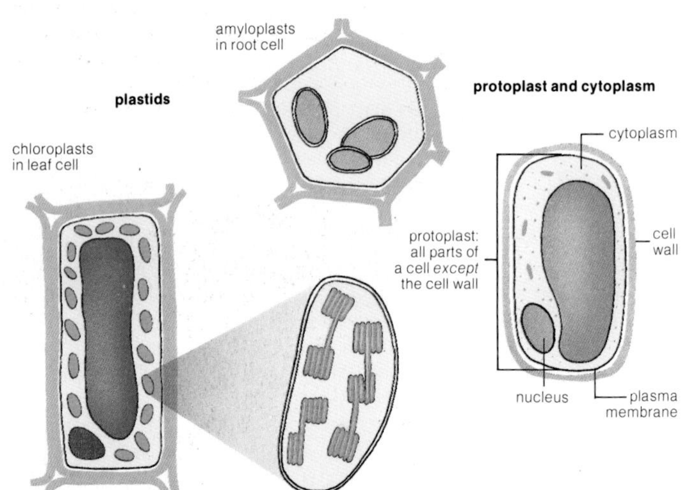

plastid (*n*) the general name for the type of organelle (p. 16) in plant cells which is surrounded by a double membrane (↑) and contains plastoglobuli (↓) and a network of internal membranes and vesicles (p. 20). There are several different kinds of plastids, each with a special function, e.g. chloroplasts (p. 32), chromoplasts (↓), amyloplasts (↓).
plastoglobuli (*n.pl.*) small, round droplets of lipid (p. 31), found in plastids (↑).

chromoplast (*n*) a plastid (↑) containing pigment (p. 36), e.g. the coloured plastids in the cells of the tissues (p. 88) of petals (p. 70) and fruits.

amyloplast (*n*) a plastid (↑) in the cortical (p. 89) cells of roots in many plants. The function of amyloplasts is to store starch (p. 30).

leucoplast (*n*) a plastid (↑) containing no pigment (p. 36). Leucoplasts may form pigments under certain conditions, e.g. leucoplasts in root cells form chlorophyll (p. 36) if exposed to light.

pore (*n*) any small hole in a surface or membrane (↑), which allows substances to pass through it, e.g. pores in nuclear membranes (↓).

nucleoplasm (*n*) the protoplasm (p. 16) inside the nucleus (↓) of a cell. The nucleoplasm contains the chromosomes (p. 46) and the nucleoli (↓).

nucleolus (*n*) small, dark body inside the nucleus (↓), which can only be seen during interphase (p. 45). It consists mostly of RNA (p. 51).

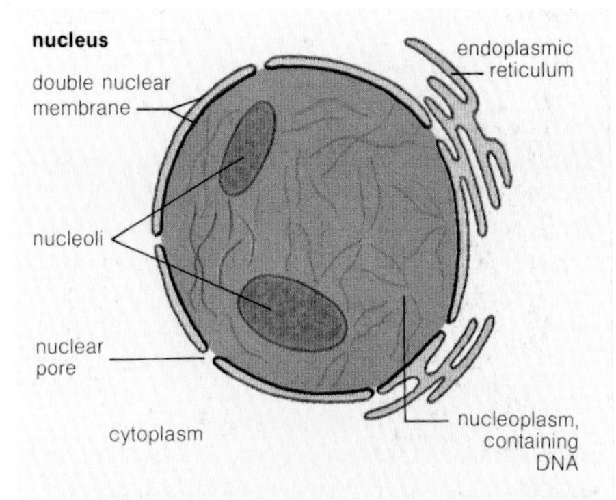

nucleus (*n*) an organelle (p. 16) of a eukaryotic (p. 16) cell, containing the nucleoplasm (↑), nucleoli (↑) and chromosomes (p. 46). Cells usually have only one nucleus, which controls most of the activities of the cell. **nuclear** (*adj*).

nuclear membrane the membrane (↑) around the nucleus (↑) of a cell. Nuclear membranes have two layers, and many pores (↑) connecting the nucleoplasm (↑) with the cytoplasm (↑).

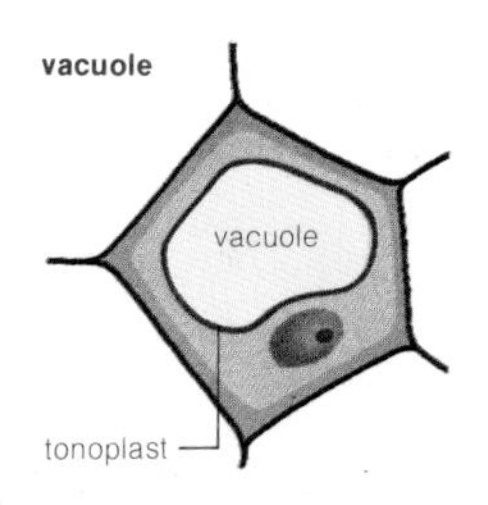

vacuole (*n*) a liquid-filled space in a cell, surrounded by a membrane (p. 18). Many plant cells, especially in leaves, have a single large vacuole and a thin layer of cytoplasm (p. 18) between it and the cell membrane (p. 18). **vacuolar** (*adj*).

tonoplast (*n*) the membrane (p. 18) enclosing the vacuole (↑) of a plant cell.

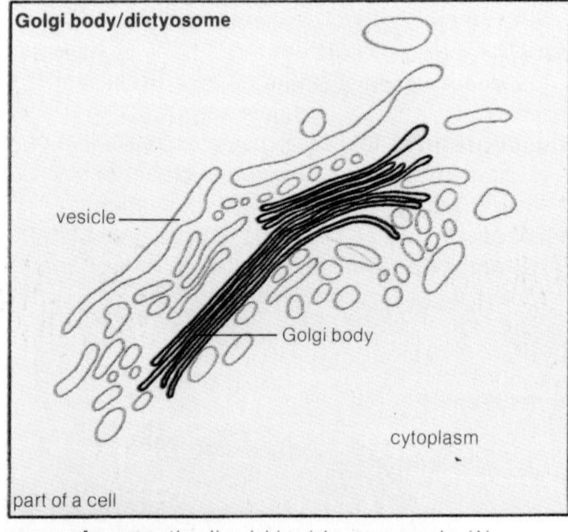

vacuolar sap the liquid inside a vacuole (↑).

Golgi body an organelle (p. 16) consisting of a group of membranes (p. 18) and vesicles (↓). The Golgi body is often important in the synthesis (p. 13) of carbohydrates (p. 28) and the secretion (p. 112) of substances, especially glycoproteins (p. 58), from the cell. In plants, it is usually called a dictyosome (↓).

dictyosome (*n*) the Golgi body (↑) of a plant cell.

vesicle (*n*) any small body in a cell or organelle (p. 16) which is surrounded by a membrane (p. 18) and contains products of metabolism (p. 14). Vesicles in the cytoplasm (p. 18) are produced mainly by the Golgi body (↑).

lysosome (*n*) an organelle (p. 16) surrounded by a membrane (p. 18), in which hydrolytic (p. 13) enzymes (p. 15) are stored. Lysosomes are common in animal cells, but probably not in plant cells.

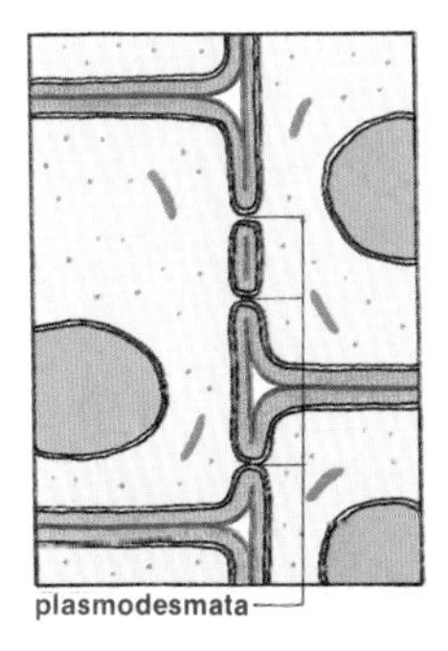

peroxisome (*n*) a small organelle (p. 16) surrounded by a membrane (p. 18), containing the enzyme (p. 15) catalase, which catalyses (p. 15) the breakdown of hydrogen peroxide (H_2O_2) to water and oxygen. Catalase prevents the build-up of H_2O_2 in cells; H_2O_2 is toxic (p. 148), and is known to be a product of some metabolic (p. 14) reactions. Peroxisomes also contain enzymes involved in the oxidation (p. 11) of glycolic acid ($COOHCH_2OH$).
plasmodesmata (*n.pl.*) threads of protoplasm (p. 16) which run through the cell wall (p. 17) between cells, along which substances can be passed. **plasmodesma** (*sing.*).

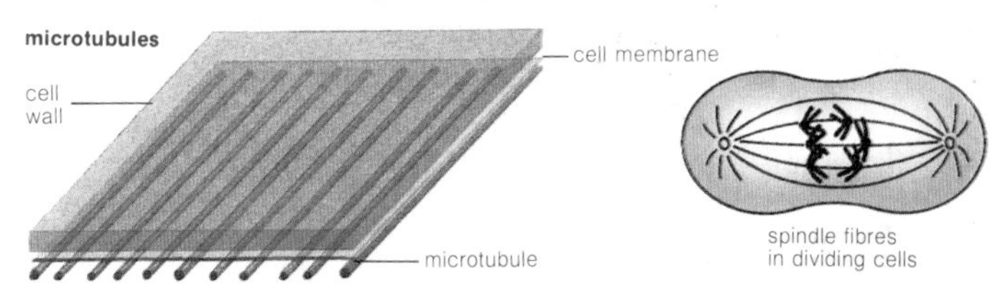

microtubule (*n*) a very thin, hollow thread of protein (p. 56) in a cell. Microtubules have several different functions. They form the spindle (p. 46) in mitosis (p. 45), control the formation of microfibrils (p. 17) in cell walls (p. 17), and form a structural part of the flagella (p. 121).
mitochondrion (*n*) a round or rod-shaped organelle (p. 16), in which the reactions of the Krebs cycle (p. 24) and the electron transfer chain (p. 40) take place. Mitochondria (*pl.*) have a smooth outer membrane (p. 18) and an inner membrane which is folded into cristae (↓).
cristae (*n.pl.*) the folds of the inner membrane (p. 18) of a mitochondrion (↑). **crista** (*sing.*).
matrix (*n*) the liquid inside a mitochondrion (↑).

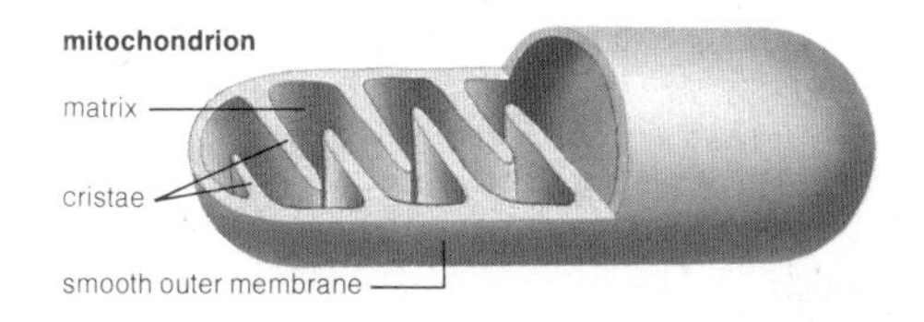

respiration (*n*) the process in which cells release energy stored in carbohydrates (p. 28) in order to drive the chemical reactions of metabolism (p. 14). Aerobic (↓) respiration involves glycolysis (↓), the Krebs cycle (p. 24), and the electron transfer chain (p. 40); this process uses oxygen, and produces carbon dioxide and ATP (p. 26). The Krebs cycle and electron transfer chain take place in the mitochondria (p. 21). Anaerobic (↓) respiration involves glycolysis (↓) and fermentation (p. 24), in which oxygen is not used. **respire** (*v*), **respiratory** (*adj*).

aerobic respiration

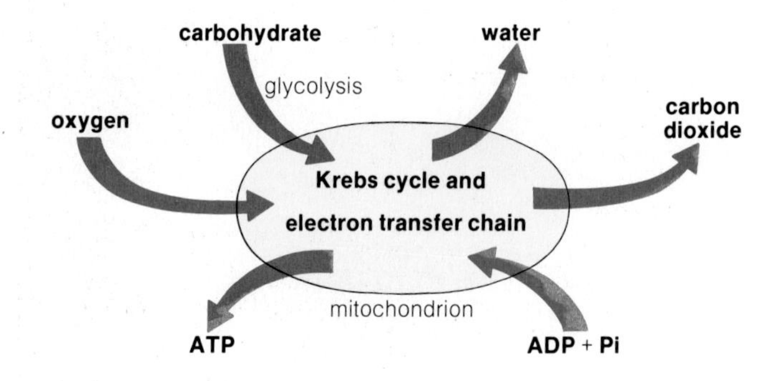

aerobic (*adj*) of respiration using molecular oxygen and involving oxidative (p. 11) processes. Also, of organisms which respire (↑) aerobically. **aerobe** (*n*).

anaerobic (*adj*) of respiration (↑) in which molecular oxygen is not used, that is, in glycolysis (↓) and fermentation (p. 24). Also, of organisms which can live without molecular oxygen, e.g. the bacteria (p. 119) that live in mud or in the gut of animals; these organisms are sometimes called anaerobes.

glycolysis (*n*) the anaerobic (↑) sequence of reactions in the breakdown of glucose (p. 28) during respiration (↑), ending with pyruvic acid (p. 24).

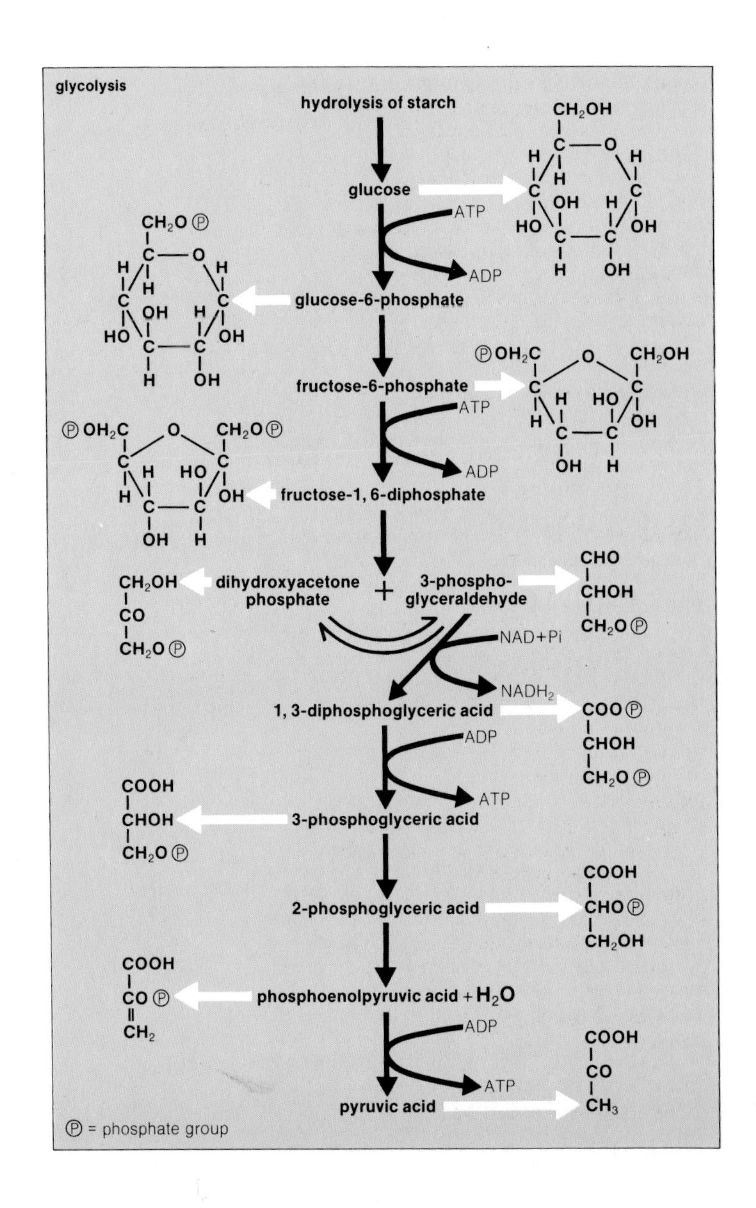
glycolysis
hydrolysis of starch
glucose
ATP
ADP
glucose-6-phosphate
fructose-6-phosphate
ATP
ADP
fructose-1, 6-diphosphate
dihydroxyacetone phosphate
+
3-phospho-glyceraldehyde
NAD+Pi
NADH2
1, 3-diphosphoglyceric acid
ADP
ATP
3-phosphoglyceric acid
2-phosphoglyceric acid
phosphoenolpyruvic acid + H2O
ADP
ATP
pyruvic acid
Ⓟ = phosphate group

fermentation (*n*) the breaking down of organic (p. 11) molecules, especially by yeast (p. 164) and bacteria (p. 119) under anaerobic (p. 22) conditions, to produce carbon dioxide and alcohol (↓) or lactic acid (↓). **ferment** (*v*).

alcohol (*n*) any one of a class of organic (p. 11) compounds with one or more hydroxyl groups (—OH), e.g. ethanol (CH_3CH_2OH).

lactic acid $CH_3CHOHCOOH$. One of the end products of fermentation (↑).

pyruvic acid $CH_3COCOOH$. The end product of glycolysis (p. 22). Pyruvic acid is the fuel for the Krebs cycle (↓) in aerobic (p. 22) organisms.

Krebs cycle a series of metabolic (p. 14) reactions in aerobic (p. 22) respiration (p. 22), in which pyruvic acid (↑) is broken down to carbon dioxide and water. The energy released in this process is used to produce ATP (p. 26) from ADP (p. 26) and orthophosphate (p. 13). The Krebs cycle takes place in the mitochondria (p. 21).

tricarboxylic acid cycle TCA cycle = Krebs cycle (↑).

citric acid cycle = Krebs cycle (↑).

GTP guanosine triphosphate. A nucleotide (p. 52) similar to ATP (p. 26), involved in the reactions of the Krebs cycle (↑).

NAD nicotinamide adenine dinucleotide. A hydrogen carrier in the Krebs cycle (↑).

FAD flavin adenine dinucleotide. A hydrogen carrier in the Krebs cycle (↑).

chemiosmosis (*n*) a process in which energy from the hydrolysis (p. 13) of ATP (p. 26) or the oxidation (p. 11) of organic (p. 11) molecules (p. 9) can be used to make an electrical and chemical gradient (↓) of protons (p. 8) across a membrane (p. 18). This gradient can be used to drive energy-requiring reactions such as the uptake of ions or the synthesis (p. 13) of ATP.

gradient (*n*) an increase or decrease of a measurable quantity over a given distance, e.g. a chemical gradient, the increase in concentration (p. 12) of a solute (p. 12) from one part of a plant to another; environmental gradient, the decrease in temperature with increasing height on a mountain.

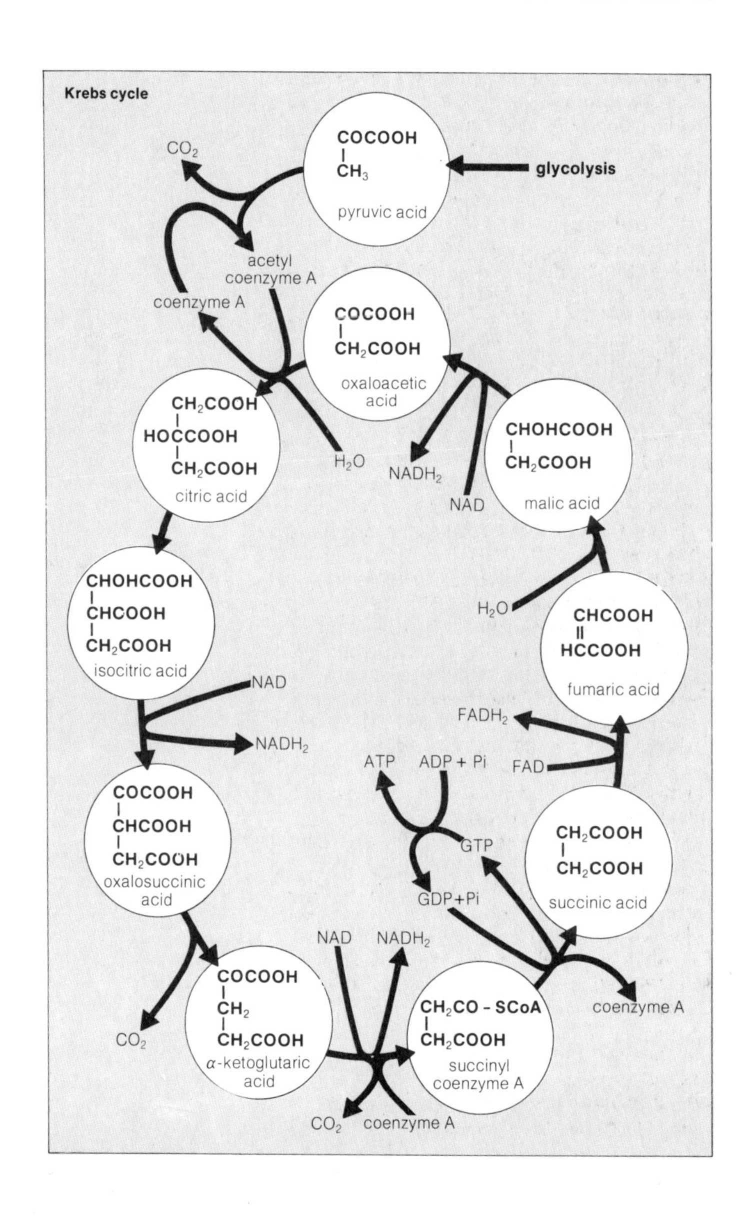

Krebs cycle
COCOOH
I
CH3
pyruvic acid
glycolysis
CO2
acetyl coenzyme A
coenzyme A
COCOOH
I
CH2COOH
oxaloacetic acid
CH2COOH
I
HOCCOOH
I
CH2COOH
citric acid
H2O
NADH2
NAD
CHOHCOOH
I
CH2COOH
malic acid
CHOHCOOH
I
CHCOOH
I
CH2COOH
isocitric acid
H2O
CHCOOH
II
HCCOOH
fumaric acid
NAD
NADH2
FADH2
ATP
ADP + Pi
FAD
COCOOH
I
CHCOOH
I
CH2COOH
oxalosuccinic acid
GTP
CH2COOH
I
CH2COOH
succinic acid
GDP+Pi
NAD
NADH2
COCOOH
I
CH2
I
CH2COOH
α-ketoglutaric acid
CH2CO - SCoA
I
CH2COOH
succinyl coenzyme A
coenzyme A
CO2
CO2
coenzyme A

photorespiration (*n*) the process in which plants, in the presence of light, high oxygen concentration (p. 12) and low carbon dioxide concentration, will take up oxygen and give off carbon dioxide, due to the oxidation of organic (p. 11) compounds produced by CO_2 fixation (p. 33). Photorespiration involves the chloroplasts (p. 32), mitochondria (p. 21) and peroxisomes (p. 21). Its function is unclear.

photorespiration

high O_2 concentration in plant tissue

low CO_2 concentration in plant tissue

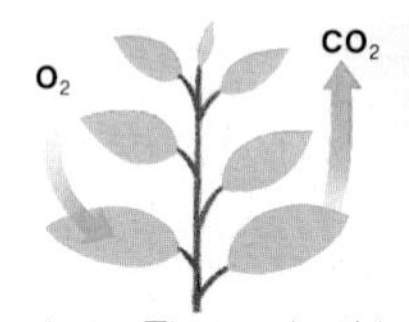

ADP adenosine diphosphate. The nucleotide (p. 52) which results from the hydrolysis (p. 13) of ATP (↓), and which, with the addition of an orthophosphate (p. 13) group, is used to synthesize (p. 13) ATP.

ATP adenosine triphosphate. The nucleotide (p. 52) which stores energy in the bonds between its three phosphate (p. 13) groups. This energy is released by hydrolysis (p. 13) to drive synthetic (p. 13) reactions in the cell.

metabolic poison any substance, e.g. cyanide, that prevents the production of ATP (↑) in cells. Without ATP as a source of energy for metabolism (p. 14), cells and organisms quickly die.

phosphorylation (*n*) the reaction in which a phosphate (p. 13) group is added to a molecule (p. 19), e.g. the phosphorylation of ADP (↑) to give ATP (↑).

oxidative phosphorylation the production of ATP (↑) from ADP (↑) and orthophosphate (p. 13), using energy released in the oxidation (p. 11) of organic (p. 11) compounds in the electron transfer chain (p. 40). Oxidative phosphorylation takes place in the mitochondria (p. 21), and is the main source of ATP in heterotrophic (p. 32) organisms. In the final reaction of the process, molecular oxygen (O_2) is reduced (p. 11) to water.

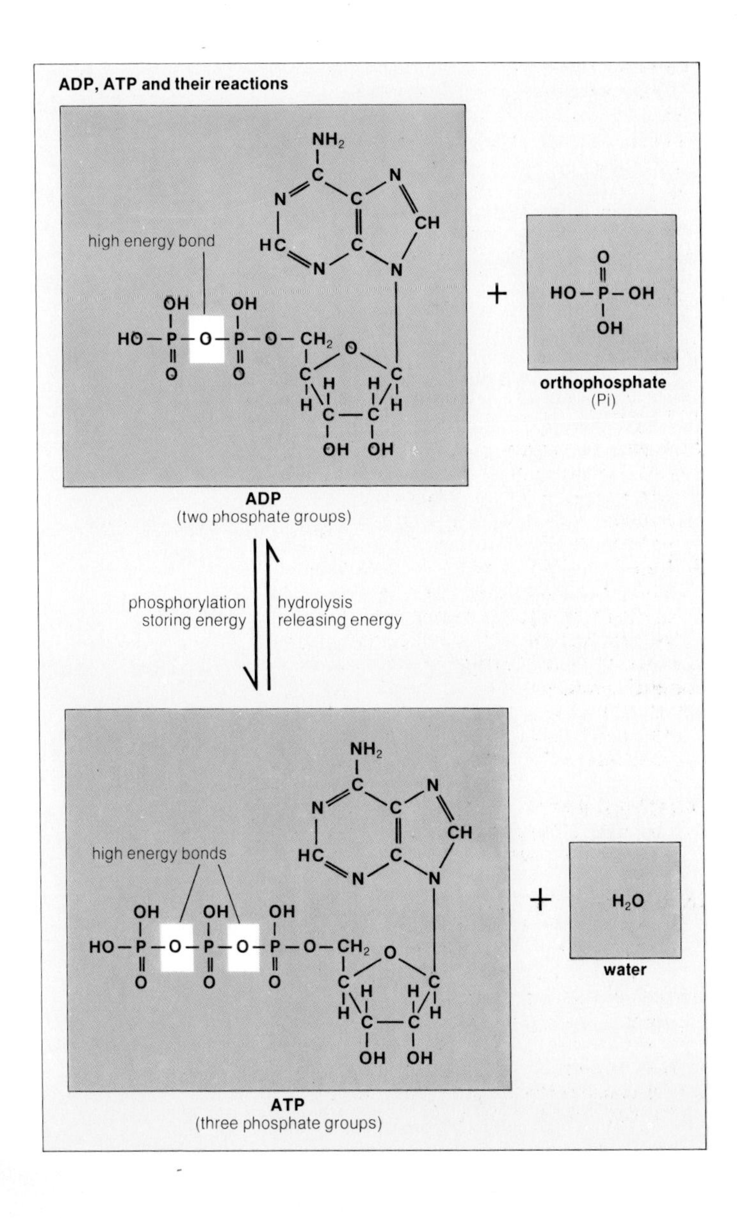
ADP, ATP and their reactions
high energy bond
NH_2
N
C
CH
HC
OH
HO — P — O — P — O — CH_2
O
H
OH OH
ADP
(two phosphate groups)
+
HO — P — OH
orthophosphate
(Pi)
phosphorylation
storing energy
hydrolysis
releasing energy
high energy bonds
HO — P — O — P — O — P — O — CH_2
ATP
(three phosphate groups)
+
H_2O
water

carbohydrate (*n*) an organic (p. 11) compound containing carbon, hydrogen and oxygen in a ratio of 1: 2: 1. Starch (p. 30) and all sugars, in which plants store the energy obtained from light during photosynthesis (p. 32), are carbohydrates.

sugar (*n*) a carbohydrate (↑) which is soluble (p. 12) in water and sweet to the taste, e.g. sucrose (↓), glucose (↓). Sugars are produced in photosynthesis (p. 32), and the energy obtained from them during respiration (p. 22) is used to drive the reactions of metabolism (p. 14).

glycoside (*n*) an organic (p. 11) compound consisting of a sugar molecule (p. 9) bonded to another organic molecule by a glycosidic bond (↓).

monosaccharide (*n*) a simple sugar, with between three and seven carbon atoms.

hexose (*n*) any monosaccharide (↑) containing six carbon atoms (p. 8), e.g. glucose (↓), fructose (↓).

pentose (*n*) any monosaccharide (↑) with five carbon atoms (p. 8). Important pentoses are ribose and deoxyribose, which are found in RNA (p. 51) and DNA (p. 51), and ribulose, which, as ribulose-diphosphate (p. 33), is used for CO_2 fixation (p. 33) in photosynthesis (p. 32).

triose (*n*) any monosaccharide (↑) with three carbon atoms (p. 8).

glucose (*n*) a hexose (↑), aldose (↓) monosaccharide (↑), with formula $C_6H_{12}O_6$. Glucose is the unit in polysaccharides (p. 30) such as starch (p. 30) and cellulose (p. 17). It is a product of photosynthesis (p. 32). With fructose (↓), it forms the disaccharide (↓) sucrose (↓).

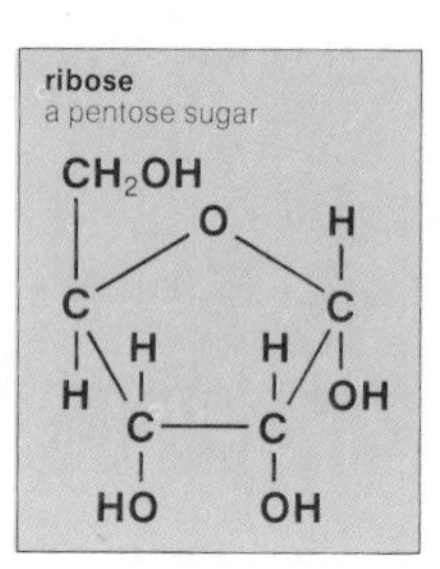

ribose
a pentose sugar

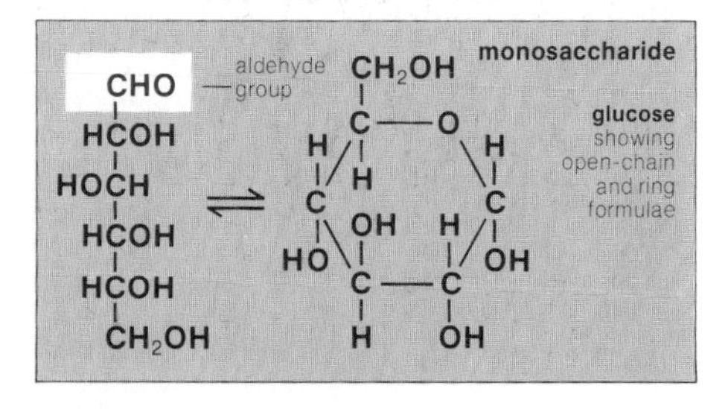

monosaccharide

glucose
showing open-chain and ring formulae

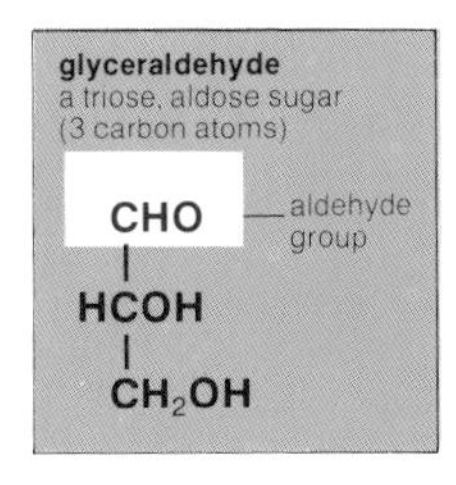

aldose (*n*) any monosaccharide (↑) in which one carbon atom is in an aldehyde (—CHO) group, e.g. glucose (↑).

ketose (*n*) any monosaccharide (↑) in which one carbon atom (p. 8) is in a ketone (—CO—) group, e.g. fructose (↓).

fructose (*n*) a hexose (↑), ketose (↑), monosaccharide (↑), with formula $C_6H_{12}O_6$. With glucose (↑), it forms the disaccharide (↓), sucrose (↓).

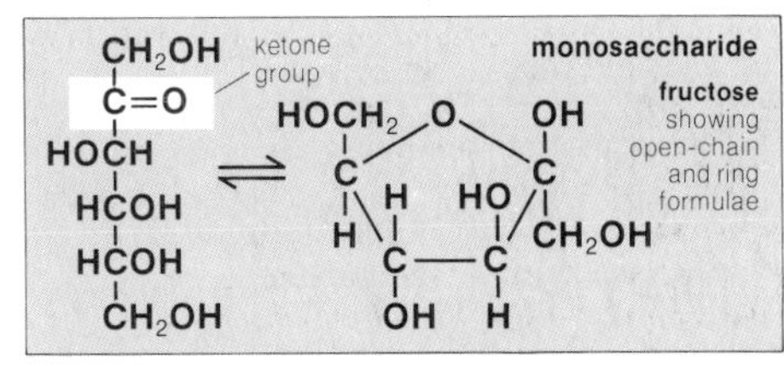

glycosidic bond the chemical bond between the sugar monomers (p. 10) in a disaccharide (↓) or polysaccharide (p. 30). The formation of the glycosidic bond is due to the reaction of the —OH group on the first carbon atom (p. 8) of one sugar molecule (p. 9) with any —OH group of another sugar molecule; H_2O is produced, and the sugars become linked by an oxygen atom.

disaccharide (*n*) a sugar made of two monosaccharide (↑) units, e.g. sucrose (↓).

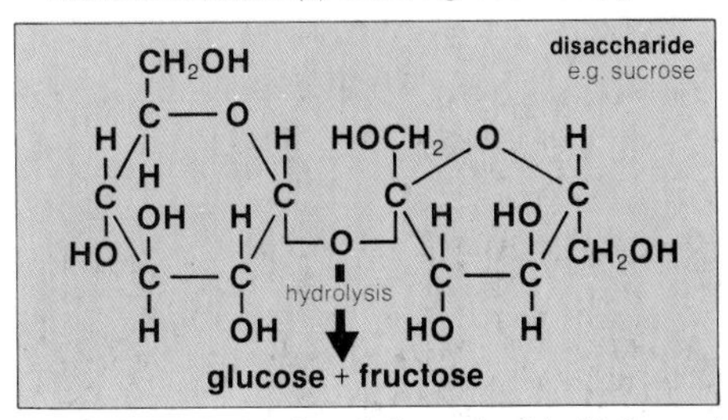

sucrose (*n*) $C_{12}H_{22}O_{11}$. A disaccharide (↑), formed from a molecule (p. 9) of glucose (↑) and a molecule of fructose (↑), found only in plants. It is the sugar which is obtained from sugar cane and sugar beet.

oligosaccharide (*n*) any sugar made of anything between two and ten monosaccharide (p. 28) units.

polysaccharide (*n*) any polymer (p. 10) consisting of many monosaccharide (p. 28) units, e.g. starch (↓), cellulose (p. 17).

polysaccharide
e.g. starch (amylopectin)

starch (*n*) a polysaccharide (↑) in which the carbohydrate (p. 28) produced during photosynthesis (p. 32) is stored in plants. Starch is a polymer (p. 10) of glucose (p. 28) units. It is deposited as small grains in chloroplasts (p. 32), and sometimes in amyloplasts (p. 19).

amylose (*n*) a form of starch (↑) made of straight chains of glucose (p. 28) monomers (p. 10).

amylopectin (*n*) a form of starch (↑) in which the glucose (p. 28) molecules (p. 9) are in branched chains.

amylase (*n*) an enzyme (p. 15) which catalyzes (p. 15) the breakdown of starch (↑) into monosaccharide (p. 28) units.

diastase (*n*) = amylase (↑).

inulin (*n*) a polysaccharide (↑) made of fructose (p. 29) monomers (p. 10). Inulin is a storage product in the roots of many plants.

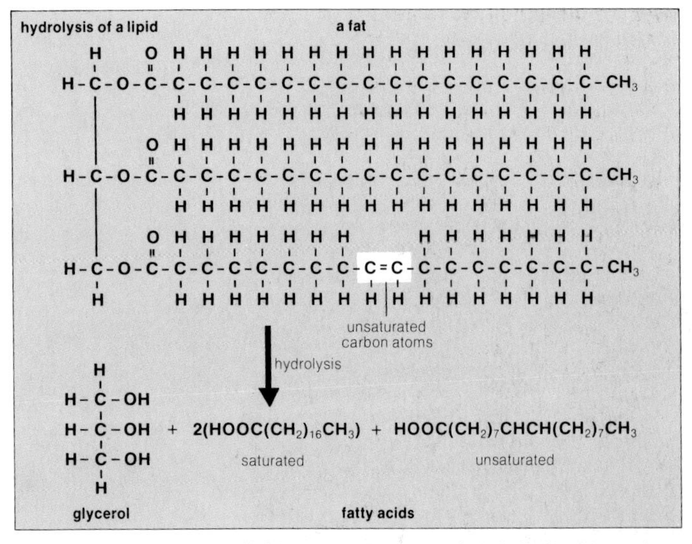

lipid (*n*) one of a group of chemical compounds, which contain glycerol (↓) and fatty acids (↓). Lipids are insoluble (p. 12) in water.

fatty acid an organic (p. 11) acid with the general formula $C_nH_{2n}O_2$. A fatty acid molecule (p. 9) is a straight chain and the number of carbon atoms (p. 8) is usually even.

saturated (*adj*) of any organic (p. 11) molecule (p. 9) with no double bonds between its carbon atoms (p. 8), e.g. the fatty acid (↑) palmitic acid, $(CH_2)_{15}COOH$.

unsaturated (*adj*) of any organic (p. 11) molecule (p. 9) with at least one double bond between carbon atoms (p. 8), e.g. the fatty acid (↑) oleic acid, $(CH_2)_8CHCH(CH_2)_7COOH$.

glycerol (*n*) $CH_2OHCHOHCH_2OH$. A compound which combines with fatty acids (↑) to form lipids (↑).

phospholipid (*n*) a lipid (↑) containing one or more phosphate (p. 13) groups.

aromatic (*adj*) of organic (p. 11) compounds in which carbon atoms (p. 8) are arranged in rings of six.

photosynthesis (*n*) the process by which plants use energy from sunlight to produce carbohydrates (p. 28) from carbon dioxide (CO_2) and water (H_2O), i.e. the conversion of simple inorganic (p. 11) compounds to complex organic (p. 11) compounds. Sunlight energy is captured by molecules (p. 9) of chlorophyll (p. 36) in the chloroplasts (↓) of cells in green leaves. The general equation for photosynthesis is $CO_2 + 4H_2O \rightarrow (CH_2O) + 3H_2O + O_2$. Some bacteria (p. 119), also use this process. **photosynthetic** (*adj*), **photosynthesize** (*v*).

autotrophic (*adj*) able to synthesize (p. 13) food from simple chemical compounds, using energy from light or chemical reactions. Most plants are autotrophic. **autotroph** (*n*).

heterotrophic (*adj*) of organisms which need a supply of organic (p. 11) matter for growth. Such organisms cannot synthesize (p. 13) organic matter using energy from light. Fungi (p. 163) and animals are heterotrophic, as are many bacteria (p. 119). **heterotroph** (*n*).

chloroplast (*n*) a green plastid (p. 18) containing chlorophyll (p. 36). Chloroplasts are the site of photosynthesis (↑). They contain their own DNA (p. 51) and reproduce (p. 59) themselves. Chloroplasts are found in the cells of leaf tissues (p. 88), and those of green stems.

chloroplast

1 to more than 100 chloroplasts per cell

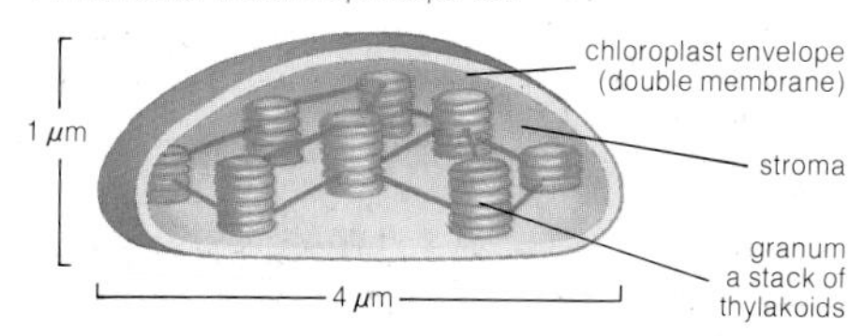

chloroplast envelope the double membrane (p. 18) surrounding the chloroplast (↑).

stroma (*n*) the parts of the chloroplast (↑) in between the stacks of grana (↓). The dark reaction (↓) of photosynthesis (↑) takes place in the stroma.

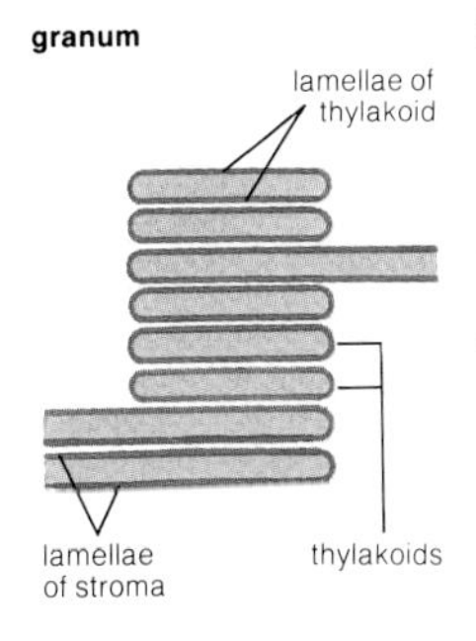

grana (*n.pl.*) the stacks of flat vesicles (p. 20) - or thylakoids (↓) - in the chloroplast (↑), where the pigments (p. 36) and enzymes (p. 15) of the light reaction (p. 36) of photosynthesis (↑) are found. **granum** (*sing.*).

lamellae (*n.pl.*) the membranes (p. 18) of the grana (↑) of chloroplasts (↑). **lamella** (*sing.*)

thylakoid (*n*) a flat vesicle (p. 20) in the grana (↑) of a chloroplast (↑).

CO_2 fixation the process in which CO_2 dissolved in the intercellular spaces (p. 95) is fixed into an organic (p. 11) molecule (p. 9) in the chloroplasts (↑) of plant cells. This is an important part of the dark reaction (↓). Usually, CO_2 reacts with ribulose-diphosphate (↓) to produce two molecules of PGA (↓).

reductive pentose pathway the set of reactions in photosynthesis (↑) in which CO_2 is fixed by ribulose-diphosphate (↓), a pentose (p. 28) sugar, to give PGA (↓), which is used to produce a hexose (p. 28) sugar and more ribulose-diphosphate, which is used for further CO_2 fixation (↑). This pathway is driven by energy from ATP (p. 26) produced in the light reaction (p. 36). It also requires $NADPH_2$ (p. 38) from the light reaction for the reduction (p. 11) of PGA.

dark reaction the part of photosynthesis (↑) controlled by enzymes (p. 15) rather than light, i.e. CO_2 fixation (↑) and the reductive pentose pathway (↑).

ribulose-diphosphate a compound consisting of a molecule (p. 9) of the pentose (p. 28) sugar ribulose and two phosphate (p. 13) groups. It is the main compound involved in CO_2 fixation (↑) during photosynthesis (↑). It is also called ribulose-bis-phosphate and RuDP.

ribulose-diphosphate carboxylase an enzyme that catalyzes (p. 15) CO_2 fixation (↑) by ribulose-diphosphate (↑).

phosphoglyceric acid PGA. A compound with three atoms (p. 8) of carbon, which is the first product of the reaction between CO_2 and ribulose-diphosphate (↑) in the reductive pentose pathway (↑) of photosynthesis (↑).

PGA = phosphoglyceric acid.

glyceric acid-3-phosphate = phosphoglyceric acid (p. 33).

Calvin cycle the reductive pentose pathway (p. 33) of photosynthesis (p. 32), named after one of its discoverers. The cycle was worked out in the 1940s and 1950s. It is also known as the C_3 pathway (↓).

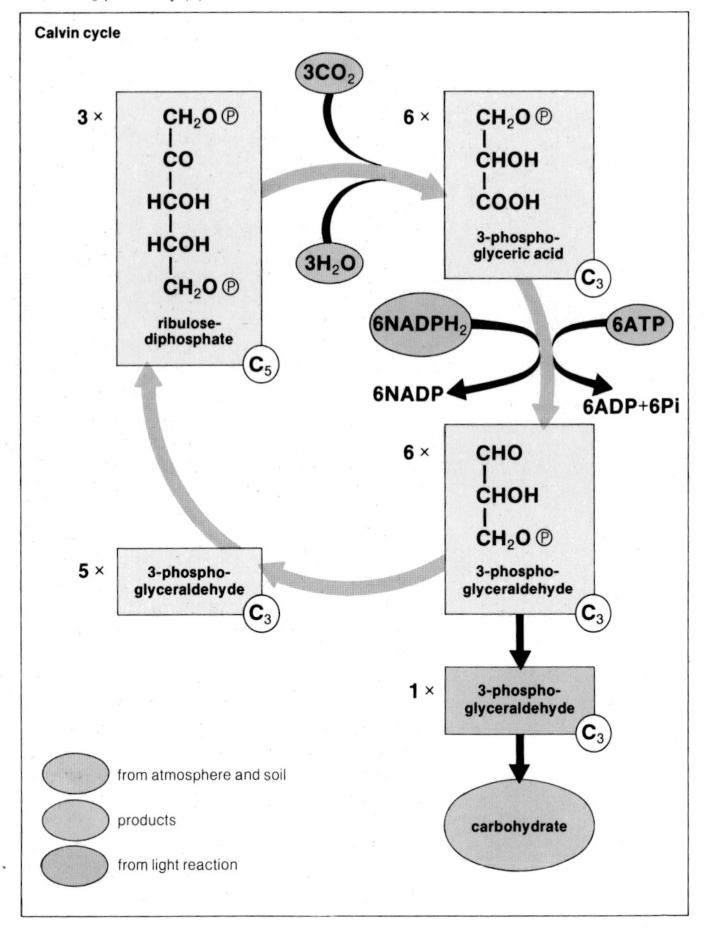

sedoheptulose (*n*) a monosaccharide (p. 28) with seven carbon atoms (p. 8), produced in the Calvin cycle (↑).

C_3 pathway the fixation of CO_2 (p. 33) by ribulose-diphosphate (p. 33) to produce two molecules (p. 9) of a compound with three carbon atoms (p. 8) (PGA (p. 33)). Most plants use this pathway, and are called C_3 plants. It is also known as the reductive pentose pathway (p. 33) and the Calvin cycle (↑).

C_4 pathway of CO_2 fixation

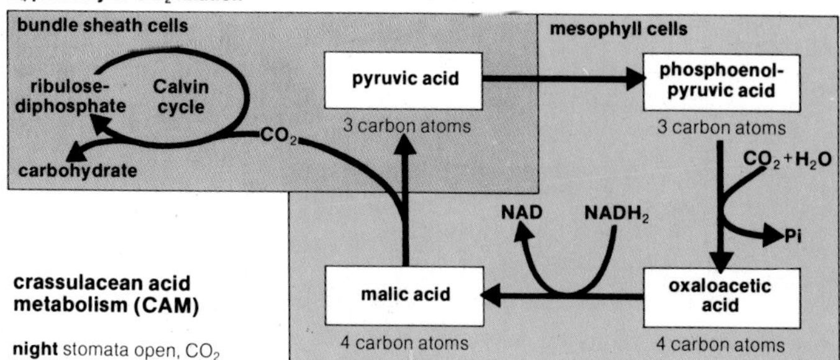

C_4 pathway kind of CO_2 fixation (p. 33), found especially in tropical (p. 162) monocotyledons (p. 130). In this pathway CO_2 is fixed by a compound with three carbon atoms (p. 8) (phosphoenolpyruvate) to produce a molecule (p. 9) with four carbon atoms (malate). This occurs in mesophyll (p. 95) cells in the leaf. The malate is then transported to bundle sheath (p. 106) cells, where the CO_2 is released and fixed by ribulose-diphosphate (p. 33) in the normal way. Plants which do this are called C_4 plants.

crassulacean acid metabolism CAM. A kind of CO_2 fixation (p. 33), found in many succulent (p. 99) plants, such as the family Crassulaceae. CO_2 is fixed by phosphoenolpyruvate, as in the C_4 pathway (↑), to produce malate; this occurs in the night, when the stomata (p. 96) are open. During the day, when the stomata are closed, the CO_2 is released and fixed again by ribulose-diphosphate (p. 33). This reduces the water loss by transpiration (p. 101) on hot days.

crassulacean acid metabolism (CAM)

night stomata open, CO_2 enters

CO_2 fixed by phosphoenol-pyruvic acid, producing malic acid

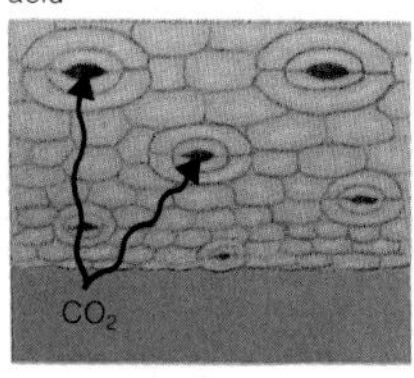

day stomata closed, preventing water loss

CO_2 released from malate and fixed by ribulose-diphosphate

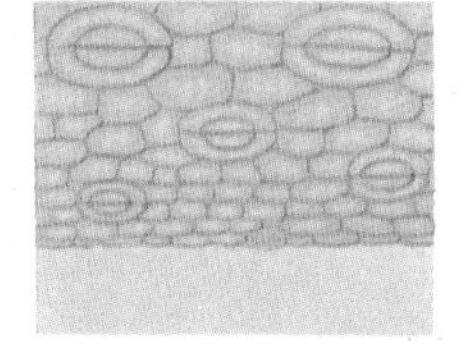

light reaction the chemical reactions of photosynthesis (p. 32) which require light. These reactions, in which pigments (↓) are used to trap light energy, are the splitting of water (H_2O) molecules (p. 9) to give hydrogen and oxygen, and the production of ATP (p. 26) and $NADPH_2$ (p. 38).

photolysis of water the part of the light reaction (↑) of photosynthesis (p. 32) in which water molecules (p. 9) are split to give hydrogen and oxygen.

Hill reaction the name for the part of the light reaction (↑) of photosynthesis (p. 32), after R. Hill who first observed it in 1937. It involves the reduction (p. 11) of NADP to $NADPH_2$ (p. 38).

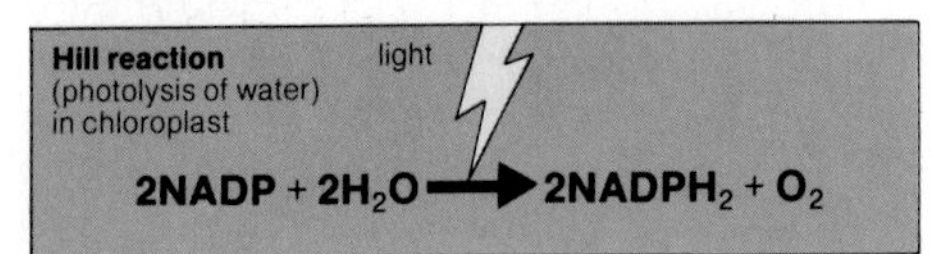

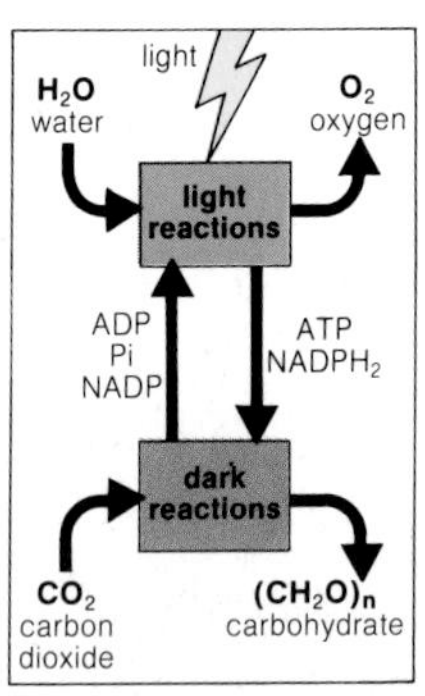

the links between the light reactions and dark reactions of photosynthesis

pigment (*n*) any coloured substance present in the tissues (p. 88) of an organism. Pigments absorb energy from light. Some of them, like chlorophyll (↓), are important in photosynthesis (p. 32). Others, e.g. phytochromes (p. 116), help to control growth.

chlorophylls (*n.pl.*) magnesium-containing green pigments (↑), found in the chloroplasts (p. 32) of all plants, which trap light energy, of blue and red wavelengths (p. 38), for photosynthesis (p. 32). Chlorophylls give plants their green colour. The two most important chlorophylls are chlorophyll *a* ($C_{55}H_{72}O_5N_4Mg$) and chlorophyll *b* ($C_{55}H_{70}O_6N_4Mg$).

accessory pigment a pigment (↑) which is involved in photosynthesis (p. 32) but not directly in the capture of sunlight energy, e.g. carotenoids (↓).

plastocyanin (*n*) a blue, copper-containing protein (p. 56), which is an electron (p. 8) carrier in the light reaction of photosynthesis (p. 32).

plastoquinone (*n*) a non-protein (p. 56) electron (p. 8) carrier in the light reaction (↑) of photosynthesis (p. 32).

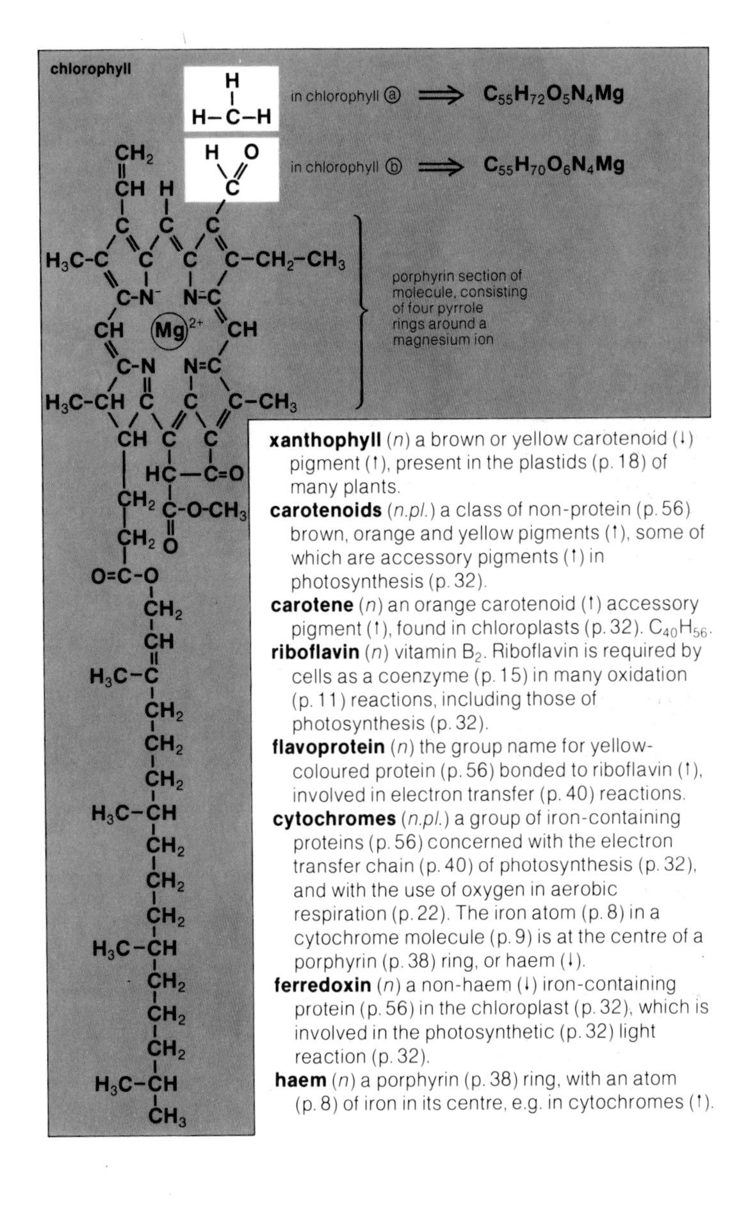

xanthophyll (*n*) a brown or yellow carotenoid (↓) pigment (↑), present in the plastids (p. 18) of many plants.

carotenoids (*n.pl.*) a class of non-protein (p. 56) brown, orange and yellow pigments (↑), some of which are accessory pigments (↑) in photosynthesis (p. 32).

carotene (*n*) an orange carotenoid (↑) accessory pigment (↑), found in chloroplasts (p. 32). $C_{40}H_{56}$.

riboflavin (*n*) vitamin B_2. Riboflavin is required by cells as a coenzyme (p. 15) in many oxidation (p. 11) reactions, including those of photosynthesis (p. 32).

flavoprotein (*n*) the group name for yellow-coloured protein (p. 56) bonded to riboflavin (↑), involved in electron transfer (p. 40) reactions.

cytochromes (*n.pl.*) a group of iron-containing proteins (p. 56) concerned with the electron transfer chain (p. 40) of photosynthesis (p. 32), and with the use of oxygen in aerobic respiration (p. 22). The iron atom (p. 8) in a cytochrome molecule (p. 9) is at the centre of a porphyrin (p. 38) ring, or haem (↓).

ferredoxin (*n*) a non-haem (↓) iron-containing protein (p. 56) in the chloroplast (p. 32), which is involved in the photosynthetic (p. 32) light reaction (p. 32).

haem (*n*) a porphyrin (p. 38) ring, with an atom (p. 8) of iron in its centre, e.g. in cytochromes (↑).

porphyrin (*n*) type of molecular structure in which four pyrrole (↓) groups are arranged in a ring around a central metal atom (p. 8). Part of the chlorophyll (p. 36) molecule (p. 9) has this structure, the metal atom being magnesium.

pyrrole (*n*) an organic (p. 11) compound with one atom (p. 8) of nitrogen and four atoms of carbon, each bonded to a hydrogen atom, arranged in a ring. Four pyrrole groups make up the porphyrin (↑) structure in chlorophyll (p. 36) and cytochromes (p. 37).

action spectrum the wavelength (↓) or wavelengths of light which activate a biochemical process. Blue and red light are needed for photosynthesis (p. 32).

absorption spectrum the wavelengths (↓) or colours of light which are absorbed by a pigment (p. 36). Plants appear green because chlorophyll (p. 36) absorbs red and blue light and reflects green light. **spectra** (*pl.*).

wavelength (*n*) the length of a wave of light. Different wavelengths have different colours and different levels of energy.

fluorescence (*n*) the fast production of light by molecules (p. 9) of pigment (p. 36), releasing the energy trapped by the pigment from a light source. Fluorescence happens about 10^{-9} seconds after the energy was trapped, at a slightly longer wavelength (↑) than the absorption spectrum (↑). **fluoresce** (*v*).

phosphorescence (*n*) the slow production of light by molecules (p. 9) of pigment (p. 36) in a semi-stable high-energy state. Phosphorescence takes place milliseconds after the trapping of the light energy, at a longer wavelength than fluorescence (↑). **phosphoresce** (*v*).

NADP nicotinamide adenine dinucleotide phosphate. A compound which can exist in oxidized (p. 11) or reduced (p. 11) forms. The reduced form is $NADPH_2$. During the light reaction (p. 36) NADP accepts the hydrogen atoms resulting from the splitting of water molecules, giving $NADPH_2$, which is then used in the reduction of CO_2 to carbohydrate (p. 28) in the dark reaction (p. 33).

action spectra and absorption spectra in photosynthesis

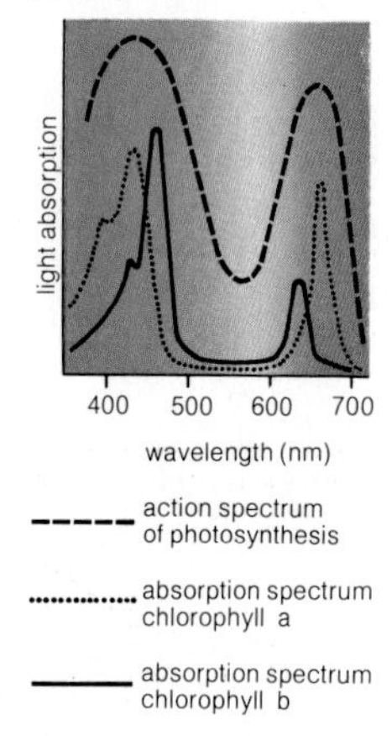

photophosphorylation (*n*) the part of the light reaction (p. 36), in which ADP (p. 26) is phosphorylated (p. 26) to ATP (p. 26) using energy from light.

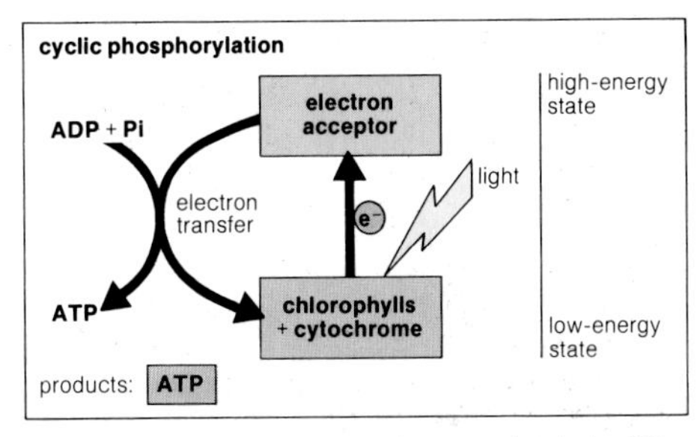

cyclic phosphorylation a photosynthetic (p. 32) reaction (p. 11) cycle in which light energy is used to produce ATP (p. 26) from ADP (p. 26) and orthophosphate (p. 13).

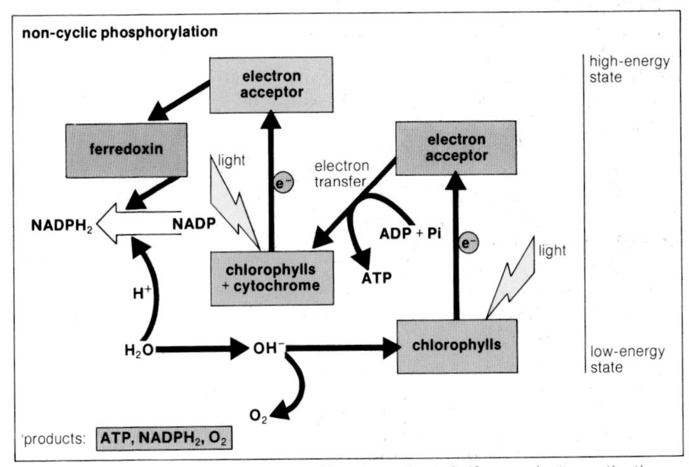

non-cyclic phosphorylation a photosynthetic (p. 32) reaction sequence in which light energy is used to produce $NADPH_2$ (↑) and oxygen from NADP and water, and ATP (p. 26) from ADP (p. 26) and orthophosphate (p. 13).

electron transfer chain (1) a set of redox (p. 11) reactions in the light reaction (p. 36) of photosynthesis (p. 32), involving plastocyanin (p. 36), plastoquinone (p. 36) and cytochromes (p. 37), in which ATP (p. 26) is produced; (2) a set of redox reactions in aerobic respiration (p. 22), involving cytochromes and also producing ATP.

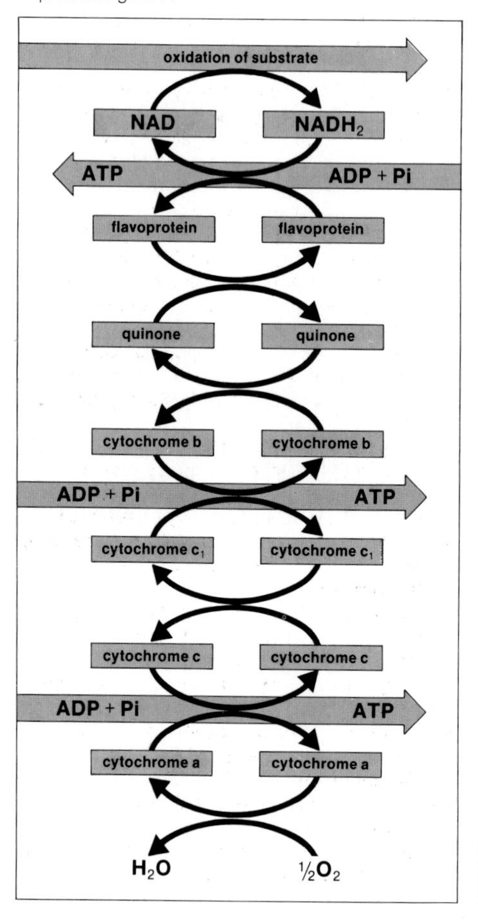

electron transfer chain
oxidative phosphorylation *via* the electron transfer chain produces three molecules of ATP

oxidized compounds

reduced compounds

genetics (*n*) the study of the mechanism of inheritance (↓), and the control of the characteristics of an organism by its genes (↓). **geneticist** (*n*).

gene (*n*) a length of DNA (p. 51) in a chromosome (p. 46). A gene codes for a particular character of the cell or organism, and can be regarded as a unit of inheritance (↓). **genetic** (*adj*).

pleiotropic (*adj*) of genes (↑) which appear to control several different characters in an organism.

genome (*n*) the genetic (↑) material on the sets of chromosomes (p. 46) in a cell. The smallest genome consists of all the genes (↑) on a haploid (p. 50) set of chromosomes. A diploid (p. 50) cell, with two sets of chromosomes, is said to have a diploid genome.

clone (*n*) a set of cells or individuals (p. 135) reproduced (p. 59) vegetatively (p. 60) from the same original cell or organism. All members of a clone have exactly the same genome (↑) or genetic (↑) material.

genotype (*n*) the allelic (p. 43) composition of a single locus (p. 44), several loci, or the entire genome (↑) of an individual (p. 135). **genotypic** (*adj*).

phenotype (*n*) the observable characteristics of an organism, regarded as the result of the interaction (p. 144) between the genotype (↑) and the environment (p. 149). **phenotypic** (*adj*).

aberration (*n*) an unusual phenotype (↑), resulting from genetic (↑) abnormality or mutation (p. 54). **aberrant** (*adj*).

genecology (*n*) the study of the distribution of genes (↑) in a population (p. 135) of organisms, in relation to their habitat (p. 149).

heredity (*n*) the passing of characters from one generation (p. 63) to the next. **hereditary** (*adj*).

inherit (*v*) to receive genetic (↑) material or characters from parents and ancestors. **inheritance** (*n*).

trait (*n*) a character or set of characters.

wild type the phenotype (↑) which most of the individuals (p. 135) in a population (p. 135) have in their natural habitat (p. 149).

Mendel's Laws the laws of inheritance (p. 41) worked out by an Austrian, Gregor Mendel, (1822-1884) in 1866. Mendel's first law is the law of segregation (↓) and the second is the law of independent assortment (↓). **Mendelian inheritance**.

segregation (*n*) the separation of each of a pair of alleles (↓) into different gametes (p. 61), as a result of meiosis (p. 49). This is the mechanism behind Mendel's first law (↑), which states that alleles brought together in the F_1 generation (↓) can be segregated in the F_2 generation (↓).

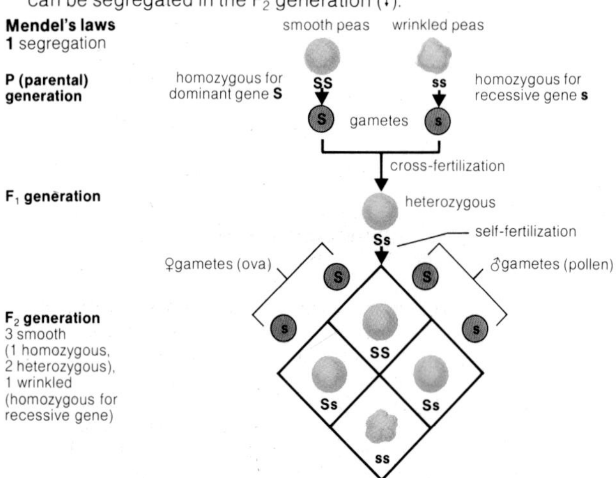

linkage (*n*) the usual inheritance (p. 41) of two or more characters together, which happens when the genes (p. 41) controlling these characters are on the same chromosome (p. 46). Linked genes can only be separated by crossing-over (p. 47) during meiosis (p. 49). Genes on the same chromosome form a linkage group.

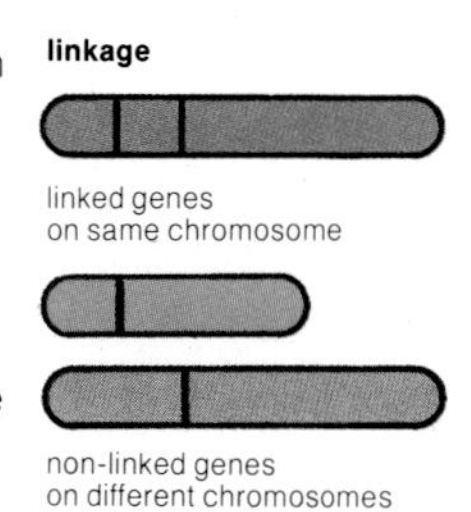

F_1 generation the first filial generation. The offspring (p. 44) of the parental generation at the beginning of a genetic (p. 41) experiment.

F_2 generation the second filial generation. The offspring (p. 44) resulting from sexual (p. 59) reproduction (p. 59) in the F_1 generation (↑).

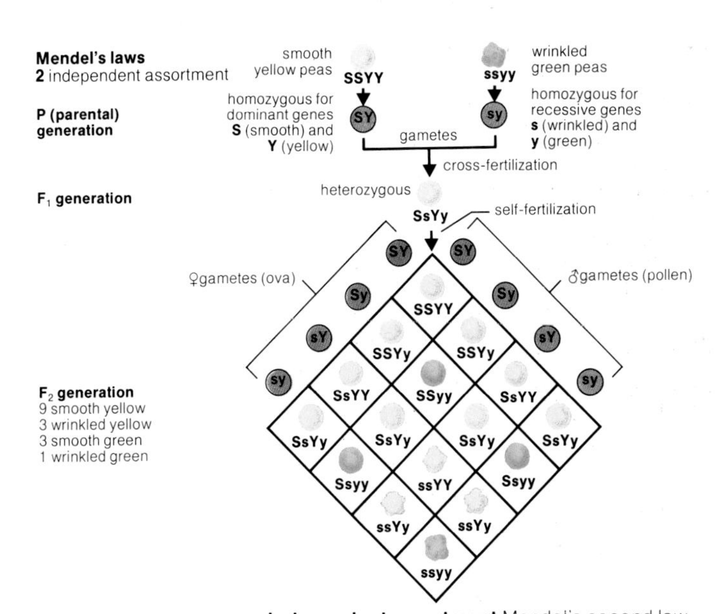

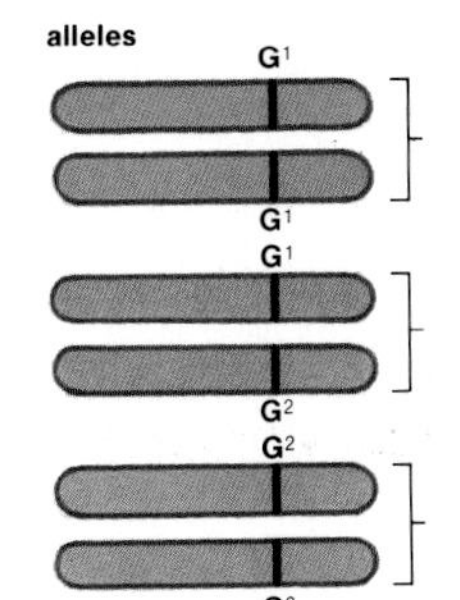

some possible combinations of 3 alleles on a chromosome pair

independent assortment Mendel's second law (↑), which states that most of the characters of parents can appear in any combination in their offspring (p. 44).

alleles (*n.pl.*) two genes (p. 41), each occupying the same position or locus (p. 44) on two homologous (p. 46) chromosomes (p. 46). Alleles may have small differences in the sequence of bases (p. 52) in their DNA (p. 51).

cytoplasmic inheritance inheritance (p. 41) of characters coded (p. 53) by DNA (p. 51) in the mitochondria (p. 21), chloroplasts (p. 32) or in other parts of the cytoplasm (p. 18).

plasmagene (*n*) any gene (p. 41) that is not contained in the nucleus (p. 19), i.e. genes that are found in the cytoplasm (p. 18). These genes are passed from one generation (p. 63) to the next by cytoplasmic inheritance (↑). The inheritance of plasmagenes does not usually obey Mendel's laws, since they are not organized into chromosomes (p. 46).

locus (*n*) a position on a chromosome (p. 46). **loci** (*pl.*).
homozygous (*adj*) having identical alleles (p. 43) at the same loci (↑) on two homologous (p. 46) chromosomes (p. 46). **homozygosity** (*n*).
heterozygous (*adj*) having non-identical alleles (p. 43) at the same loci (↑) on two homologous (p. 46) chromosomes (p. 46). **heterozygosity** (*n*).
dominant[1] (*adj*) of alleles (p. 43) which have the same effects in the heterozygous (↑) and homozygous (↑) conditions. **dominance** (*n*).
recessive (*adj*) of alleles (p. 43) whose effects can only be seen in the homozygous (↑) condition. When in the heterozygous (↑) condition, it is the dominant (↑) allele and not the recessive allele which controls the phenotype (p. 41).
isolation (*n*) the separation of one object from another, or the inability of two substances or organisms to mix with each other. Reproductive (p. 59) isolation is the inability of two or more populations (p. 135) to breed with each other, e.g. because they live in different places or different habitats (p. 149), because they flower at different times of year, or because they have different genomes (p. 41). **isolated** (*adj*).
deme (*n*) any population (p. 135) of organisms that is genetically isolated (↑) from other populations. Individuals (p. 135) in a deme breed amongst themselves, and there is no input of genetic material from other demes.
gene pool all the different genes (p. 41) present in a population (p. 135).
monohybrid inheritance the inheritance (p. 41) of one pair of genes (p. 41).
dihybrid inheritance the inheritance (p. 41) of two pairs of genes (p. 41).
pure line a series of generations (p. 63) that are homozygous (↑) for all characters.
offspring (*n*) = progeny (p. 59).
chimaera (*n*) a plant whose tissues (p. 88) are of more than one genetic (p. 41) kind. This can happen due to mutations (p. 54) in a cell of a very young plant, or can be caused by grafting (p. 69).

locus

homologous chromosomes

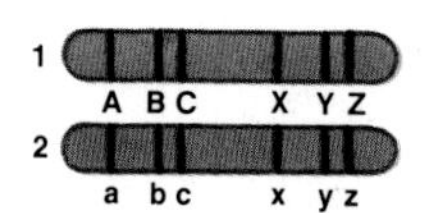

alleles **A**, **B**, **C**, **X**, **Y**, **Z**, occupy same loci (positions) on chromosome 1 as alleles **a**, **b**, **c**, **x**, **y**, **z**, on chromosome 2

homozygosity and heterozygosity

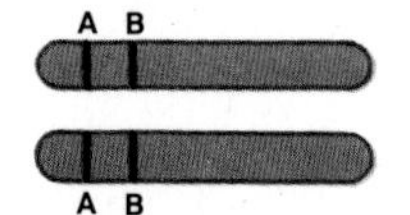

chromosome pair homozygous for gene **A** and gene **B**

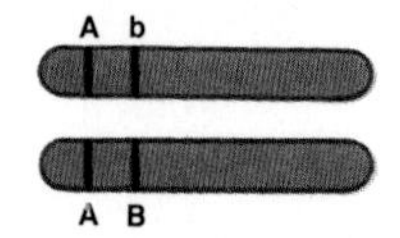

chromosome pair homozygous for gene **A**, heterozygous for gene **B**

mitosis
(only two pairs of homologous chromosomes shown for clarity)

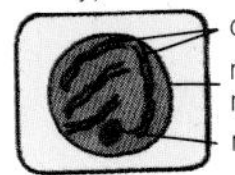

prophase chromosomes become visible in the nucleus, each one split into two chromatids, joined at the centromere.

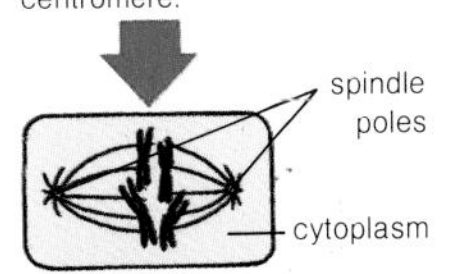

metaphase nuclear membrane and nucleolus have disintegrated. Spindle fibres form. Chromosomes shorter and thicker, arranged midway between the spindle poles.

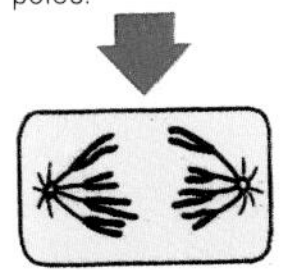

anaphase chromatids separate at centromeres. Sister chromatids drawn to opposite poles of the spindle.

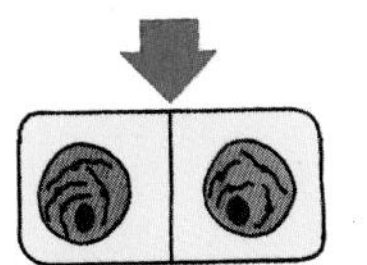

telophase nuclear membranes and nucleoli re-form. Chromosomes begin to lose their compact structure. The new cell wall is laid down.

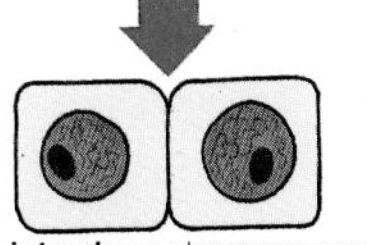

interphase chromosomes no longer visible.

cell division the process in which a cell divides to form two new cells, each containing a nucleus (p. 19). Cell division is either mitotic (↓) or meiotic (p. 49).

cytokinesis (*n*) the division of a cell into two cells.

binary fission the division of a cell into two identical cells.

mitosis (*n*) vegetative (p. 60) or somatic (↓) cell division (↑), in which the chromosomes (p. 46) in the nucleus (p. 19) are duplicated into two chromatids (p. 46). The nuclear membrane (p. 19) breaks down, the centromeres (p. 46) divide, and the chromatids move to either end of the cell on the spindle (p. 46). Nuclear membranes re-form around each group of chromatids, and a new cell wall (p. 17) is laid down between them. In this way each new cell gains exactly the same chromosomes and genetic (p. 41) material. The four stages of mitosis are prophase (↓), metaphase (↓), anaphase (↓) and telophase (↓). **mitotic** (*adj*).

somatic (*adj*) of any process, or part of an organism, that is not connected with sexual reproduction (p. 59), e.g. mitosis (↑) is somatic cell division (↑).

prophase (*n*) the first stage in cell division (↑), when the chromosomes (p. 46) in the nucleus (p. 19), when stained (p. 171), become visible as short, thick, helically (p. 51) coiled threads.

metaphase (*n*) the second stage in cell division (↑), in which the nuclear membrane (p. 19) breaks down, and the centromeres (p. 46) of the chromosomes (p. 46) position themselves at the centre of the spindle (p. 46), forming the metaphase plate.

anaphase (*n*) the third stage of cell division (↑), in which the chromatids (p. 46) separate from each other and move to the ends, or poles, of the spindle (p. 46).

telophase (*n*) the last stage in cell division (↑), when the chromatids (p. 46) are at either end of the spindle (p. 46) and the new nuclear membranes (p. 19) are being synthesized (p. 13).

interphase (*n*) the period between one cell division (↑) and the next.

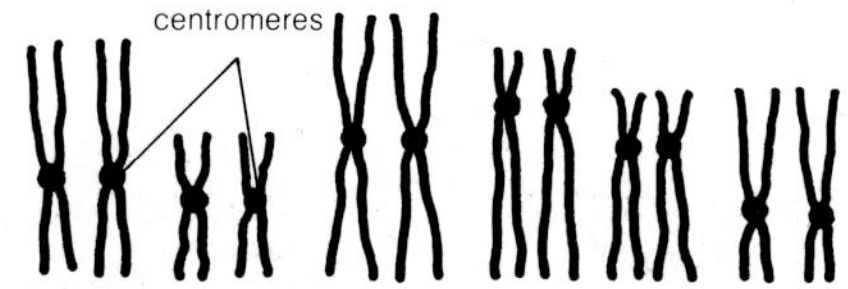

pairs of homologous chromosomes

chromosome (*n*) threadlike bodies containing DNA (p. 51), RNA (p. 51), and protein (p. 56), found in the nuclei (p. 19) of all cells. They are usually only visible during cell division (p. 45), during which they become shorter and thicker. All the vegetative (p. 60) cells in a plant, and in a species (p. 134), have the same number of chromosomes.

homologous (*adj*) of two similar chromosomes (↑) which pair with each other during meiosis (p. 49). Homologous chromosomes have identical sequences of loci (p. 44). Members of a pair of homologous chromosomes have centromeres (↓) in the same position and arms of the same length as each other.

chromatid (*n*) one of the pair of strands which result from the duplication of a chromosome (↑) during prophase (p. 45) and metaphase (p. 45).

centromere (*n*) the point of attachment of a chromosome (↑) to the spindle (↓) during cell division (p. 45).

spindle (*n*) the minute curved threads of protein (p. 56) which appear during cell division (p. 45), spreading across the cell from either end. The movement of chromosomes (↑) during cell division is organized on the spindle. The protein threads of the spindle are microtubules (p. 21), which are formed during metaphase (p. 45).

centriole (*n*) a small granule outside the nuclear membrane (p. 19), which divides at mitosis (p. 45), forming the two ends of the spindle (↑). Centrioles are found in all animal cells, but in plants they can only be seen in motile (p. 121) male gametes (p. 61).

centrosome (*n*) the region in the cytoplasm (p. 18) that gives rise to centrioles (↑) during cell division (p. 45).

the effect of colchicine in mitosis

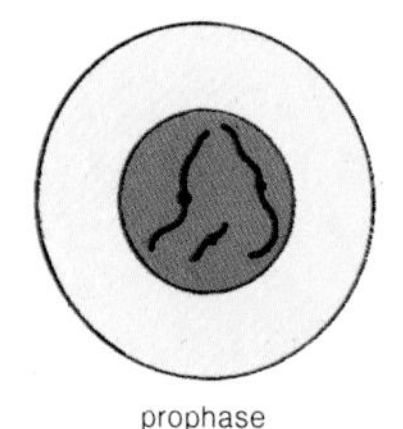
prophase nucleus

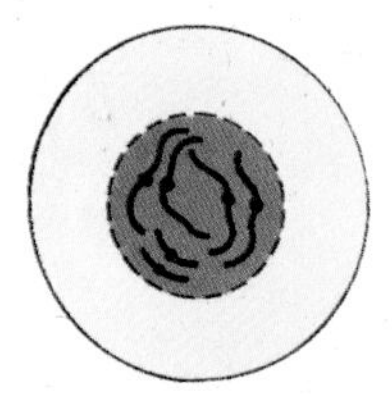
chromosomes divide normally into chromatids, no spindle formed

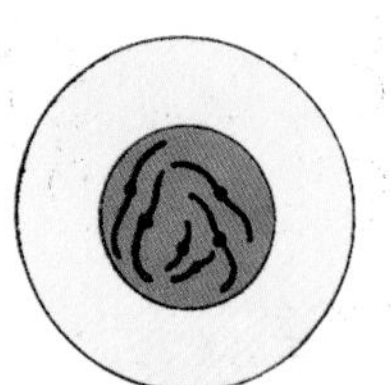
nuclear membrane reforms, containing twice as many chromosomes as original nucleus

crossing-over
during first meiotic division

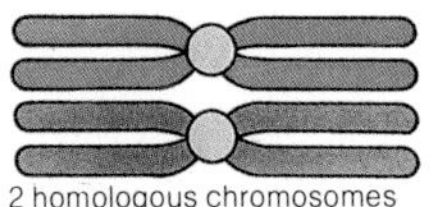
2 homologous chromosomes

chiasma

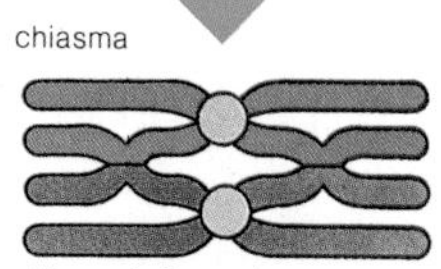
chiasmata formed

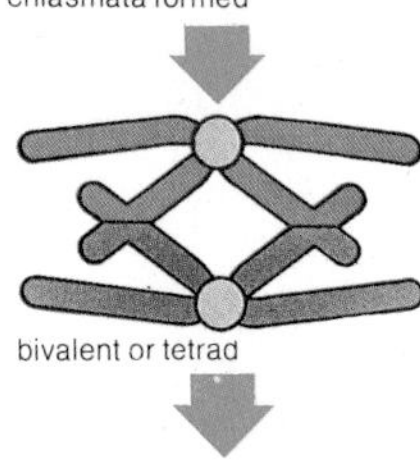
bivalent or tetrad

genetic material exchanged, chromosomes separate

colchicine (*n*) an alkaloid (p. 148) which inhibits (p. 14) the synthesis (p. 13) of the spindle (↑) in the metaphase (p. 45) stage of mitosis (p. 45). It is used in many genetic (p. 41) experiments, as its effect is to produce tetraploid (p. 50) cells when the new nuclear membrane (p. 19) forms at the end of mitosis.

crossing-over the exchange of some of the corresponding parts, each with the same set of loci (p. 44), of homologous (↑) chromosomes (↑) at the first meiotic (p. 49) division, during which chiasmata (↓) are formed. The result of crossing-over is recombination (↓).

synapsis (*n*) the pairing of homologous (↑) chromosomes (↑) during zygotene (p. 49) in the first meiotic (p. 49) prophase (p. 45).

chiasmata (*n.pl.*) the points in a bivalent (↓) where the two chromosomes (↑) appear to be joined and crossed over. Chiasmata can be observed during diplotene (p. 49). **chiasma** (*sing.*).

bivalent (*n*) a pair of joined homologous (↑) chromosomes (↑) during the first meiotic (p. 49) prophase (p. 45).

tetrad[1] (*n*) another name for a bivalent (↑) in meiosis (p. 49), so called because it consists of four chromatids (↑) in two pairs.

recombination (*n*) the process by which offspring (p. 44) can gain combinations of genes (p. 41) different from the combinations in either of their parents. This happens as a result of crossing-over (↑) of chromosomes (↑).

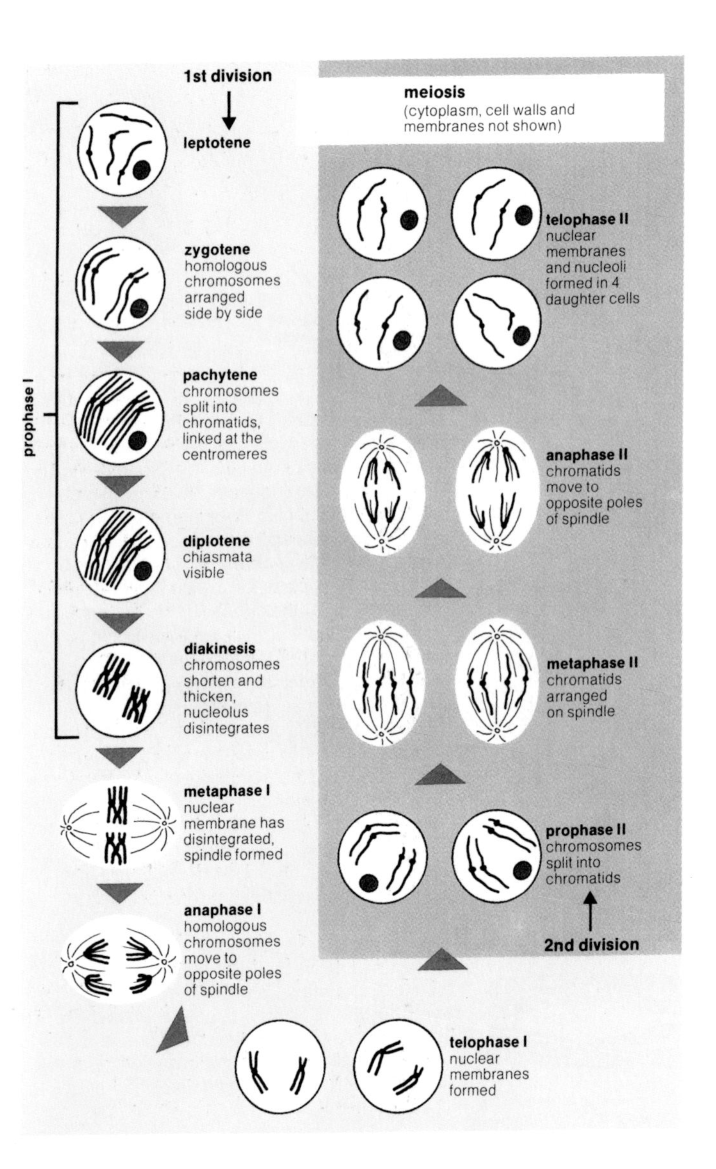
1st division
meiosis
(cytoplasm, cell walls and membranes not shown)
prophase I
leptotene
zygotene
homologous chromosomes arranged side by side
pachytene
chromosomes split into chromatids, linked at the centromeres
diplotene
chiasmata visible
diakinesis
chromosomes shorten and thicken, nucleolus disintegrates
metaphase I
nuclear membrane has disintegrated, spindle formed
anaphase I
homologous chromosomes move to opposite poles of spindle
telophase I
nuclear membranes formed
prophase II
chromosomes split into chromatids
2nd division
metaphase II
chromatids arranged on spindle
anaphase II
chromatids move to opposite poles of spindle
telophase II
nuclear membranes and nucleoli formed in 4 daughter cells

mitosis	meiosis
no pairing of homologous chromosomes	pairing of homologous chromosomes
splitting of chromatids at centromere	no splitting of chromatids at centromere until 2nd prophase
daughter nuclei have *same* number of chromosomes as parent nucleus	daughter nuclei have *half* the number of chromosomes as parent nuclei
2 daughter nuclei produced	4 daughter nuclei produced

differences between mitosis and meiosis

meiosis (*n*) cell division (p. 45) that produces haploid (p. 50) sex cells (p. 61) from diploid (p. 50) cells. Meiosis involves two cell divisions: (1) the replicated homologous (p. 46) chromosomes (p. 46) pair with each other on the spindle (p. 46); crossing-over (p. 47) happens at this stage. The chromosomes then separate to either end of the spindle. (2) The chromatids of each chromosome come apart at the centromere (p. 46), and separate to each end of the second spindle. There is usually no interphase (p. 45) between the two divisions. Meiosis occurs in all organisms which reproduce (p. 59) sexually (p. 59). **meiotic** (*adj*).

reduction division a name sometimes given to meiosis (↑), because the daughter cells receive a haploid (p. 50) set of chromosomes (p. 46) from the diploid (p. 50) parent cell.

leptotene (*n*) the first stage in the first meiotic (↑) prophase (p. 45), in which the chromosomes (p. 46) appear as thin threads.

zygotene (*n*) the stage in the first meiotic (↑) prophase (p. 45) when the homologous (p. 46) chromosomes (p. 46) come together to form bivalents (p. 47).

pachytene (*n*) the stage of the first meiotic (↑) prophase (p. 45), when the chromosomes (p. 46) become shorter and thicker, and can clearly be seen to have replicated (p. 54) into chromatids (p. 46).

diplotene (*n*) the stage in the first meiotic (↑) prophase (p. 45), when the centromeres (p. 46) of paired chromosomes (p. 46) move away from each other and crossing-over (p. 47) can be seen.

diakinesis (*n*) the last stage in the first meiotic (↑) prophase (p. 45), when the chromosomes (p. 46) are shortest and thickest, and the nuclear membrane (p. 19) disappears.

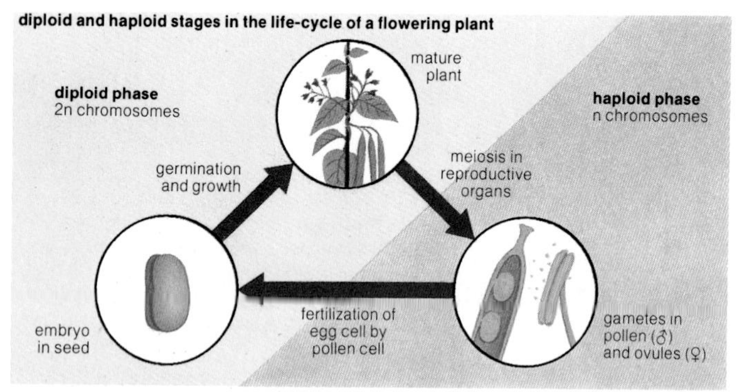

haploid (*adj*) of cells with one set of chromosomes (p. 46) in their nuclei (p. 19).

diploid (*adj*) of cells with two sets of chromosomes (p. 46) in their nuclei (p. 19). The sets are said to be homologous (p. 46).

triploid (*adj*) of cells which have three sets of homologous (p. 46) chromosomes (p. 46) in their nuclei (p. 19).

tetraploid (*adj*) of cells which have four sets of homologous (p. 46) chromosomes (p. 46) in their nuclei (p. 19).

triploid
e.g. formation of endosperm in angiosperms

♂ haploid gamete (n)

diploid endosperm cell (2n) in embryo sac

fusion

triploid endosperm mothercell (3n)

mitotic divisions

endosperm

polyploid (*adj*) of cells which have three or more sets of homologous (p. 46) chromosomes (p. 46) in their nuclei (p. 19).

allopolyploid (*n*) a polyploid (↑) species (p. 134) with sets of chromosomes (p. 46) from two or more different species. This can be the result of hybridization (p. 63) between species.

autopolyploid (*n*) a polyploid (↑) species (p. 134) with all sets of chromosomes (p. 46) coming from the same species.

nucleic acid a long chain polymer (p. 10) consisting of nucleotide (p. 52) units. There are two kinds of nucleic acid, DNA (↓) and RNA (↓), which are found in the cells of all living organisms.

DNA deoxyribonucleic acid. The main nucleic acid (↑) in the chromosomes (p. 46) of the nucleus (p. 19) of a cell. The DNA molecule consists of two chains of nucleotide (p. 52) polymer (p. 10), arranged in a double helix (↓). The sugar in the nucleotides of DNA is deoxyribose. DNA controls protein synthesis (p. 57) by the processes of transcription (p. 56) and translation (p. 56). It is replicated (p. 54) by a self-copying process, and it is the hereditary (p. 41) material of all cellular (p. 169) organisms and some viruses (p. 118).

diagram of the DNA double helix

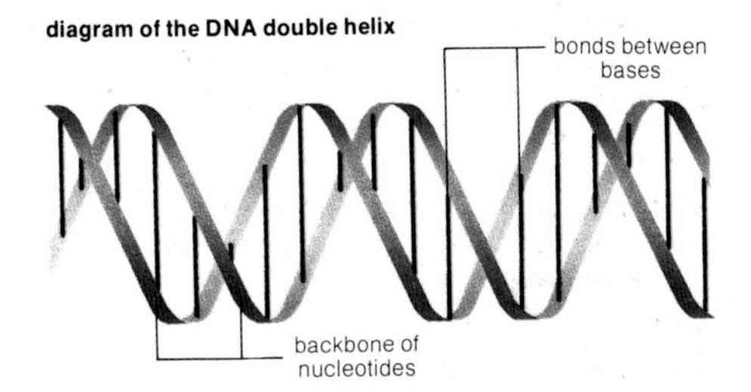

helix (*n*) a thread or line coiled like a screw. Molecules of DNA (↑) have this shape, with two helices (*pl.*) coiled together. **double helix**.

RNA ribonucleic acid. The nucleic acid (↑) directly involved in protein synthesis (p. 57). RNA differs from DNA (↑) by having uracil (p. 53) instead of thymine (p. 53) and ribose instead of deoxyribose in its nucleotides (p. 52). The RNA polymer (p. 10) is usually a single strand. There are three main kinds of RNA: *messenger RNA* (mRNA), which carries the genetic code (p. 54) from the nucleus to the cytoplasm (p. 18); *transfer RNA* (tRNA), to which amino acids (p. 56) are attached before protein synthesis (p. 57); and *ribosomal RNA* (rRNA), which is a structural part of the ribosomes (p. 56).

the common bases in the nucleotides of DNA and RNA

	purines	pyrimidines
DNA only		thymine
DNA and RNA	adenine guanine	cytosine
RNA only		uracil

nucleotide (*n*) a molecule with a pentose (p. 28) sugar, a phosphate (p. 13) group, and a purine (↓) or pyrimidine (↓) base (↓) containing nitrogen. Nucleotides are the units which form the long chain polymers (p. 10), nucleic acids (p. 51).

base[2] (*n*) a purine (↓) or pyrimidine (↓) unit.

nucleotide basic structure

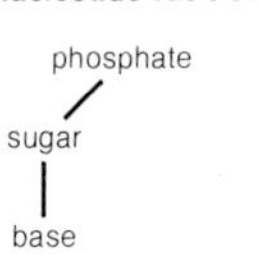

different nucleotides have different sugars and different bases

a chain of nucleotides

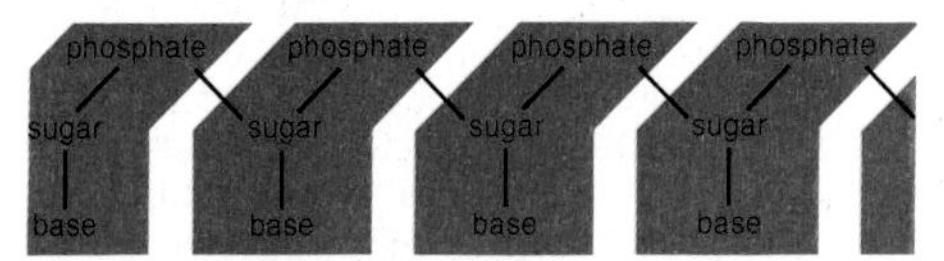

purine (*n*) one of the two kinds of nitrogen-containing base (↑) in nucleic acids (p. 51). A purine molecule (p. 9) consists of two rings of carbon and nitrogen atoms (p. 8). The main purines in nucleic acids are adenine (↓) and guanine (↓).

adenine (*n*) a purine (↑) base (↑) which pairs with thymine (↓) in DNA (p. 51) and with uracil (↓) in RNA (p. 51).

guanine (*n*) a purine (↑) base (↑) which pairs with cytosine (↓) in DNA (p. 51) and RNA (p. 51).

pyrimidine (*n*) one of the two kinds of nitrogen-containing base (↑) found in nucleic acids (p. 51). The molecules (p. 9) consist of a single ring of carbon and nitrogen atoms (p. 8). The main pyrimidines in nucleic acids are thymine (↓), cytosine (↓), and uracil (↓).

cytosine (*n*) a pyrimidine (↑) base (↑), which pairs with guanine (↑) in DNA (p. 51) and RNA (p. 51).

thymine (*n*) a pyrimidine (↑) base (↑), which pairs with adenine (↑) in DNA (p. 51).

uracil (*n*) a pyrimidine (↑) base (↑) found in RNA (p. 51), which pairs with adenine (↑) during transcription (p. 56) and translation (p. 56).

codon (*n*) a sequence of three nitrogen-containing bases (↑) in the triplet code (p. 54) on a molecule (p. 9) of messenger RNA (p. 51), which pairs with an anticodon (↓) on a molecule of transfer RNA (p. 51) during translation (p. 56). Because each base only pairs with one other base, each codon has its own anticodon, e.g. a codon consisting of adenine (↑), guanine (↑) and cytosine (↑), or AGC, will pair with an anticodon of uracil (↑), cytosine and guanine, or UCG.

anticodon (*n*) a sequence of three nitrogen-containing bases (↑) in a molecule (p. 9) of transfer RNA (p. 51), which pairs with a codon (↑) on a molecule of messenger RNA (p. 51) during translation (p. 56). Each molecule of transfer RNA has only one anticodon, corresponding to the particular amino acid (p. 56) to which it is attached during protein synthesis (p. 57), e.g. one of the anticodons for the amino acid serine is a sequence consisting of uracil (↑), cytosine (↑) and guanine (↑), or UCG.

nonsense codon a codon (↑) that does not code for any amino acid (p. 56). Only three of the 64 codons in the genetic code (p. 54) are nonsense, and their function is to code for the ends of polypeptide (p. 56) chains.

genetic code the name given to the 64 possible sequences in which any three of the four nitrogen-containing bases (p. 52) of RNA (p. 51), adenine (p. 53), uracil (p. 53), guanine (p. 53) and cytosine (p. 53), can be arranged. Each group of three, or triplet, codes for a particular amino acid (p. 56) in protein synthesis (p. 57). Because there are only 20 amino acids, most of them are coded for by more than one triplet of bases.

triplet code a name for the genetic code (↑), so called because the code consists of nitrogen-containing bases (p. 52) in groups of three.

replication (*n*) the process by which new DNA (p. 51) is made. The two strands in the DNA double helix (p. 51) separate, and a new strand of nucleotide (p. 52) polymer (p. 10) is synthesized (p. 13) on each one. Because each nitrogen-containing base (p. 52) in the nucleotide units of the polymer will only pair with one other base, the new DNA has exactly the same sequence of bases as the old. This self-copying process is the basis of heredity (p. 41). **replicate** (*v*).

mutation (*n*) the general term for a change in the sequence of nucleotides (p. 52) in the DNA (p. 51) of a cell e.g. the replacement of a pair of nitrogen-containing bases (p. 52) in the DNA chain by another pair, the turning round of a sequence of nucleotides in the chromosome (p. 46), or the loss of a whole chromosome or piece of DNA. Mutations can be inherited (p. 41) if they are present in the gametes (p. 61). They can be harmless, useful, or can cause death, depending on where they occur in a chromosome. They occur rarely, but this rate can be speeded up by mutagens (↓). Mutations result in variation (p. 135) between individuals (p. 135), and the natural selection (p. 140) of this variation leads to evolution (p. 139). **mutate** (*v*).

mutagen (*n*) a factor that causes mutations (↑), e.g. X-rays, gamma rays or certain chemicals. **mutagenic** (*adj*).

mutant (*n*) an individual (p. 135) which shows the effects of a mutation (↑) phenotypically (p. 41).

amino acids and the genetic code

amino acid general formula

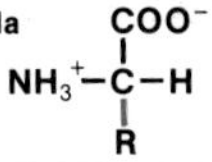

$$NH_3^+-\underset{R}{\overset{COO^-}{C}}-H$$

R = side group

codon	amino acid	side group (R)	side group (R)	amino acid	codon
AAA, AAG	lysine	$-CH_2CH_2CH_2CH_2NH_3^+$	$-H$	glycine	GGU, GGC, GGA, GGG
AAU, AAC	asparagine	$-CH_2CONH_2$	$-CH_2COO^-$	aspartic acid	GAU, GAC
ACU, ACC, ACA, ACG	threonine	$-CHOHCH_3$	$-CH_2CH_2COO^-$	glutamic acid	GAA, GAG
AGU, AGC	serine	$-CH_2OH$	$-CH_3$	alanine	GCU, GCC, GCA, GCG
AGA, AGG	arginine	$-CH_2CH_2CH_2NHC(NH_2)=N^+H_2$	CH_3CHCH_3 (bond at CH)	valine	GUU, GUC, GUA, GUG
AUU, AUC, AUA	isoleucine	$CH_3CH_2CHCH_3$ (bond at CH)	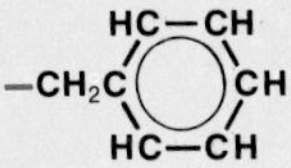 $-CH_2C$ (ring: HC—CH, CH, HC—CH)	phenylalanine	UUU, UUC
AUG	methionine	$-CH_2CH_2SCH_3$	$-CH_2CH(CH_3)CH_3$	leucine	UUA, UUG
CCU, CCC, CCA, CCG	proline	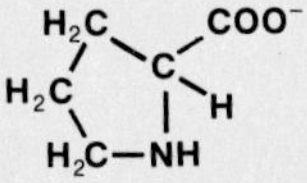 ring: H_2C—$C(COO^-)(H)$—NH—H_2C—H_2C	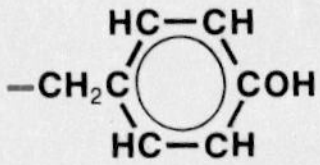 $-CH_2C$ (ring: HC—CH, COH, HC—CH)	tyrosine	UAU, UAC
CAU, CAC	histidine	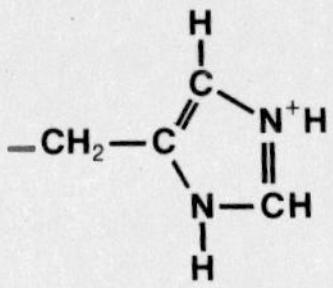 $-CH_2-C$ (ring: C(H)=, N^+H, =CH, N(H)—CH)		NONSENSE	UAA, UAG
CAA, CAG	glutamine	$-CH_2CH_2CONH_2$	$-CH_2SH$	cysteine	UGU, UGC
CGU, CGC, CGA, CGG	arginine	$-CH_2CH_2CH_2NHC(=N^+H_2)NH_2$	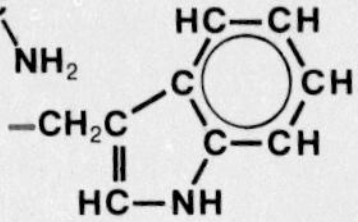 $-CH_2C$ (ring: C, HC—CH, CH, C—CH; HC—NH)	tryptophan	UGG
CUU, CUC, CUA, CUG	leucine	$-CH_2CH(CH_3)CH_3$		NONSENSE	UGA
			$-CH_2OH$	serine	UCU, UCC, UCA, UCG

protein (*n*) a substance made of one or more polypeptides (↓), which themselves are made of amino acids (↓). There are very many different kinds of protein, each with its own sequence of amino acids. Some are structural, e.g. in membranes (p. 18), and others are enzymes (p. 15) which catalyze (p. 15) reactions in cells.

amino acid any one of a class of organic (p. 11) compounds with a carboxyl group (—COOH), an amino group (NH_2) and a 'side-group' all attached to a central carbon atom (p. 8). Different amino acids have different side-groups. There are about 20 different amino acids found in proteins (↑), which they form when linked together in a chain or polymer (p. 10).

peptide (*n*) a compound made of two or more amino acids (↑) joined in a polymer (p. 10).

polypeptide (*n*) a peptide (↑) with a large number of amino acids (↑). Polypeptide chains become folded to form proteins (↑).

transcription (*n*) the process in which mRNA (p. 51) is produced in the nucleus (p. 19) of a cell, carrying in its sequence of nitrogen-containing bases (p. 52) the genetic code (p. 54) of the DNA (p. 51) in the nucleus.

translation (*n*) the part of protein synthesis (↓) in which molecules of tRNA (p. 51) carrying amino acids (↑) are matched with the genetic code (p. 54) carried by mRNA (p. 51), so that the amino acids are joined together in the correct sequence to make a polypeptide (↑). This takes place on the ribosomes (↓).

endoplasmic reticulum a system of membranes (p. 18) in the cytoplasm (p. 18), where much of protein synthesis (↓) takes place. Endoplasmic reticulum may be rough (with ribosomes (↓)), or smooth (without ribosomes).

ribosome (*n*) small body made of rRNA (p. 51) and protein (↑). It is the site of protein synthesis (↓) and the process of translation (↑). Cells can contain many thousands of ribosomes, which are found either on the endoplasmic reticulum (↑) or as polysomes (↓).

polysome (*n*) a group of ribosomes (↑) joined together by a strand of mRNA (p. 51).

peptide bond between amino acids

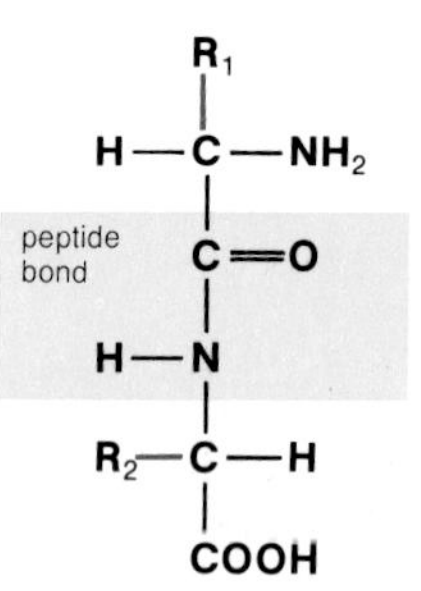

R_1 and R_2 are side groups

protein synthesis the building of polymers (p. 10) of amino acids (↑), which takes place on the ribosomes (↑) of a cell. Each amino acid is attached to a molecule (p. 9) of tRNA (p. 51) before synthesis starts. The anticodon (p. 53) on the tRNA must be matched with the codon (p. 53) on a molecule of mRNA (p. 51) which runs through the ribosome, before its amino acid can be joined onto the polypeptide (↑) chain which will become the protein (↑).

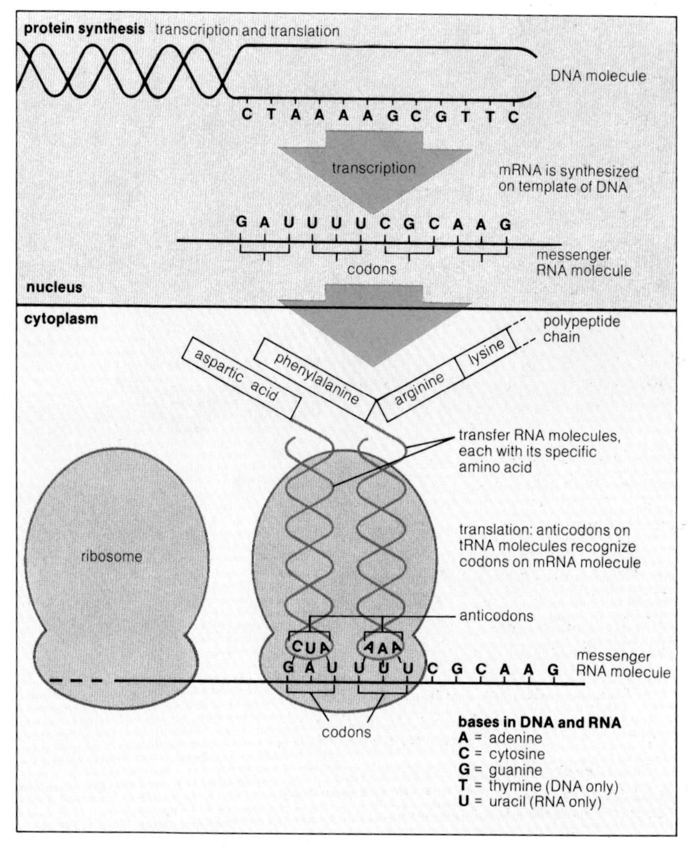

primary, secondary, tertiary and quaternary structure of proteins

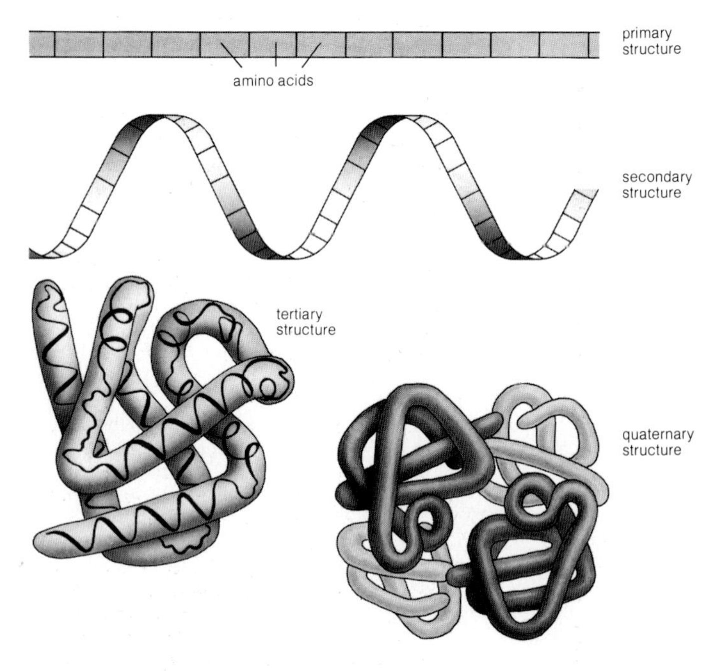

protein structure the structure of proteins (p. 56) can be studied at four levels - primary, secondary, tertiary and quaternary. The *primary structure of a protein* is the sequence of amino acids (p. 56) in a polypeptide (p. 56). The *secondary structure of a protein* is the coiling of the polypeptide into a helix (p. 51) or pleated sheet. The *tertiary structure of a protein* is the twisting and folding of the polypeptide helix or pleated sheet to form a three-dimensional protein molecule (p. 9). The *quaternary structure of a protein* is the structure of several protein molecules when bonded together.

glycoprotein (*n*) a protein (p. 56) bonded to a sugar.

reproduction (*n*) the process in which an organism produces offspring (p. 44) like itself. Reproduction can be sexual (↓) or asexual (↓), and is one of the most important characteristics of living organisms. **reproduce** (*v*), **reproductive** (*adj*).
progeny (*n*) the offspring (p. 44) or young produced by an organism during reproduction (↑).
sexual (*adj*) of reproduction (↑) which involves the fusion (p. 61) of two cells and their nuclei (p. 19) from two parent individuals (p. 135), so that the offspring (p. 44) receives genetic (p. 41) material from both parents. Sexual reproduction occurs in all the divisions (p. 134) of the plant kingdom (p. 134). **sex** (*n*).
breed (*v*) to reproduce (↑) sexually (↑).
asexual (*adj*) of reproduction (↑) from one individual (p. 135), without the fusion (p. 61) of sex cells (p. 61) from two different parents. Asexual reproduction is common in the plant kingdom (p. 134). Many plant species can reproduce both sexually (↑) and asexually.
apomixis (*n*) the production of propagules (↓) by the female reproductive organs (p. 88) of a plant, without the sexual (↑) fusion (p. 61) of cells. In one type of apomixis, the embryo (p. 85) develops from the unfertilized haploid (p. 50) egg-cell (p. 61), in which case the offspring (p. 44) are usually sterile (p. 62). In others, the embryo develops from diploid (p. 50) tissue (p. 88) in the ovule (p. 78), in which case the offspring are fertile (p. 62). **apomictic** (*adj*).
propagule (*n*) any reproductive (↑) unit which gives rise to a new individual (p. 135), e.g. a seed, a spore (p. 66).
agamospermy (*n*) asexual (↑) production of embryos (p. 85) and seeds in flowering plants (p. 130).
apogamy (*n*) asexual (↑) reproduction (↑) in which embryos (p. 85) and propagules (↑) are produced without meiosis (p. 49) occurring.
apospory (*n*) production of a diploid (p. 50) gametophyte (p. 65) from vegetative (p. 60) cells of the sporophyte (p. 65), that is, without the production of spores (p. 66).

vegetative reproduction a type of asexual (p. 59) reproduction (p. 59) in which a whole new plant is produced from an organ (p. 88), e.g. a rhizome (↓), bulb (↓), or tuber (↓), which is not involved in sexual (p. 59) reproduction.

vegetative (*adj*) of any part of a plant which is not involved in sexual (p. 59) reproduction (p. 59). Stems, leaves and roots are vegetative organs (p. 88).

bulb (*n*) an organ (p. 88) of perennation (p. 117) and vegetative reproduction (↑) in many monocotyledons. Bulbs are usually underground, and consist of a short axis (p. 92), with many overlapping thick leaves. These leaves generally lack chlorophyll (p. 36) and contain stored food.

bulbil (*n*) a small bulb (↑).

corm (*n*) a thickened stem base, usually underground, with buds (p. 110) in the axils (p. 96) of dead leaf bases. Corms are organs (p. 88) of vegetative reproduction (↑) and perennation (p. 117).

rhizome (*n*) a stem which grows along under the ground, bearing buds (p. 110) which produce shoots. This is a way of vegetative reproduction (↑) and perennation (p. 117), as the shoots grow into whole new plants. **rhizomatous** (*adj*).

stolon (*n*) a stem which grows along the ground, producing at its nodes (p. 90) new plants with roots and upright stems. **stoloniferous** (*adj*).

runner (*n*) a stolon (↑) which produces roots and a new plant at its apex (p. 90). This is a way of vegetative reproduction (↑). After the new plant has started to grow, the runner dies and decays.

sucker (*n*) a new shoot which develops from the base of a plant or from its roots. This is a way of vegetative reproduction (↑).

tiller (*n*) a single new plant growing from the base of an old plant, especially in grasses.

tuber (*n*) a thick underground stem in which food is stored. Tubers have buds (p. 110) in modified leaf axils (p. 96), from which new plants can grow, e.g. potato. Tubers are organs (p. 88) of perennation (p. 117) and vegetative reproduction (↑).

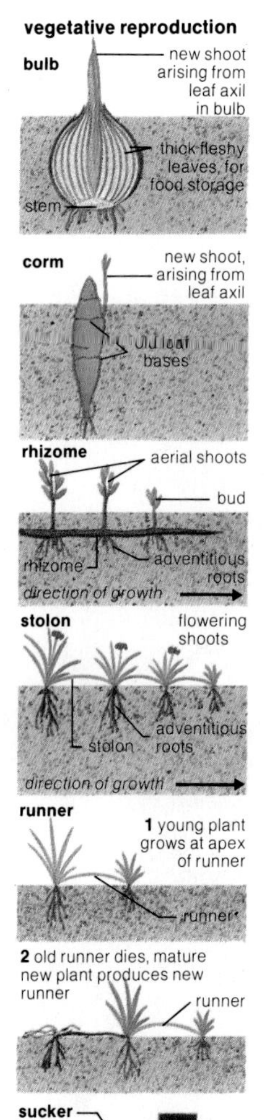

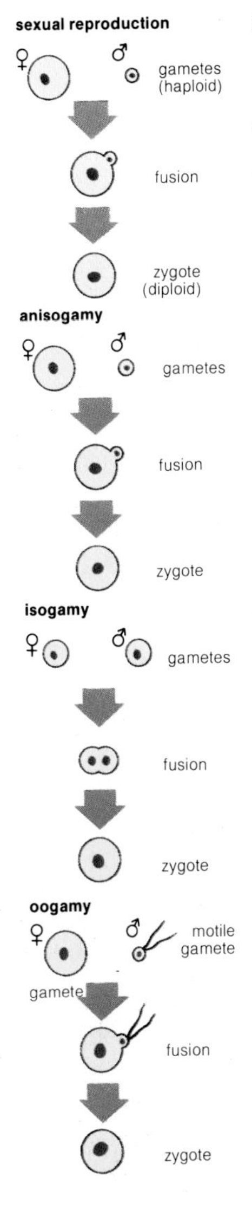

gamete (*n*) a haploid (p. 50) sex cell (↓), whose function is to join with a gamete of the opposite sex, to form a diploid (p. 50) zygote (↓). In plants, gametes are produced by the gametophyte (p. 65).

sex cell = a gamete (↑).

ovum (*n*) an egg-cell (↓) or female gamete (↑).

egg-cell (*n*) the female gamete (↑) or ovum (↑).

anisogamous (*adj*) of plants which produce gametes (↑) of different sizes, sometimes known as microgametes (male) and megagametes (female). All plants growing on land are anisogamous. **anisogamy** (*n*).

heterogamous (*adj*) having male and female gametes (↑) of different sizes, i.e. anisogamous (↑). **heterogamy** (*n*).

isogamous (*adj*) having male and female gametes (↑) of the same size. This is characteristic of some algae (p. 119). **isogamy** (*n*).

oogamous (*adj*) having a small motile (p. 121) male gamete (↑) and a large non-motile female gamete, as in bryophytes (p. 122) and pteridophytes (p. 126). **oogamy** (*n*).

zygote (*n*) a diploid (p. 50) cell which is produced by the fusion (↓) of two haploid (p. 50) gametes (↑). A fertilized (p. 62) ovum (↑) is a zygote. In plants, the zygote develops first into an embryo (p. 85) and then into a sporophyte (p. 65).

fusion (*n*) the joining together of two gametes (↑) to form a zygote (↑). Fusion can mean the joining of the cells, the joining of the nuclei (p. 19), or both. **fuse** (*v*).

conjugation (*n*) the joining together of two similar cells, usually male and female, in some algae (p. 119). **conjugate** (*v*).

gender (*n*) the sex of an individual (p. 135). Gender can be male, female, or neuter (↓).

female (*adj*) of individuals (p. 135), tissues (p. 88), organs (p. 88), etc. producing egg-cells (↑). **female** (*n*).

male (*adj*) of individuals (p. 135), organs (p. 88), tissues (p. 88), etc. producing the gametes (↑) which fertilize (p. 62) egg-cells (↑) produced by females. **male** (*n*).

neuter (*adj*) neither male nor female.

bisexual (*adj*) of organisms with male and female reproductive (p. 59) organs (p. 88) on the same individual (p. 135).

fertile (*adj*) of organisms which produce offspring, or of reproductive (p. 59) organs (p. 88) which produce viable (↓) gametes (p. 61). **fertility** (*n*).

viable (*adj*) able to carry out its function, e.g. the ability of a dormant (p. 117) seed to germinate (p. 87) when conditions become suitable. **viability** (*n*).

sterile (*adj*) of organisms which cannot produce offspring (p. 44), or of reproductive (p. 59) organs (p. 88) which do not produce gametes (p. 61), e.g. staminodes (p. 73). **sterility** (*n*).

fertilization in angiosperms

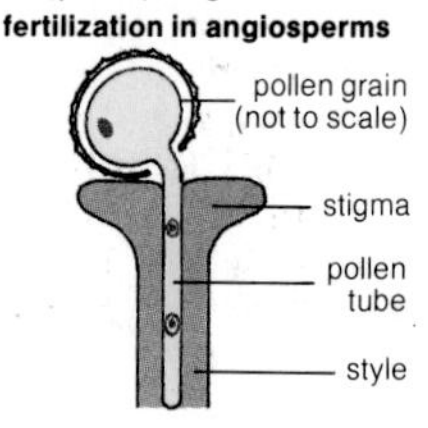

1 pollen grain lands on stigma, pollen tube grows through tissues of style carrying the male gametes

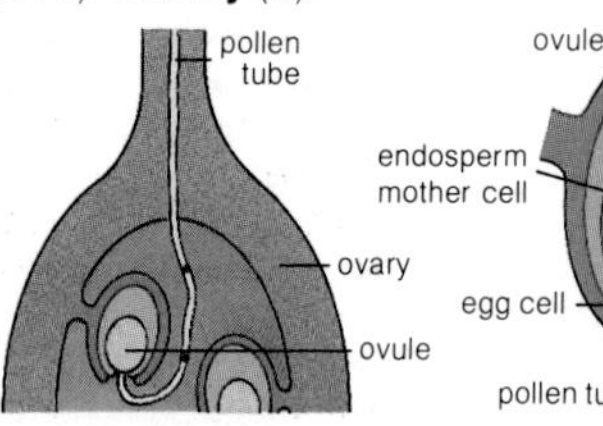

2 pollen tube grows through ovary wall and into micropyle of ovule

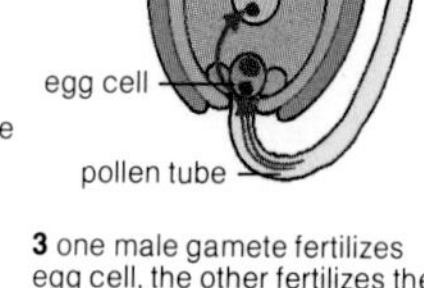

3 one male gamete fertilizes egg cell, the other fertilizes the endosperm nucleus forming endosperm mother cell

fertilization (*n*) the fusion (p. 61) of a male gamete (p. 61) with a female gamete to form a zygote (p. 61). **fertilize** (*v*).

self-fertilization (*n*) the fertilization (↑) of a female gamete (p. 61) by a male gamete from the same individual (p. 135). This is sometimes called selfing.

autogamy (*n*) = self-fertilization (↑). **autogamous** (*adj*).

cleistogamy (*n*) self-fertilization (↑) before a flower opens. The flowers of some species (p. 134) never fully open, and such species are habitually cleistogamous.

cross-fertilization (*n*) fertilization (↑) of a female gamete (p. 61) of one plant by a male gamete from another.

allogamy (*n*) the production of zygotes (p. 61) by cross-fertilization (↑). **allogamous** (*adj*).

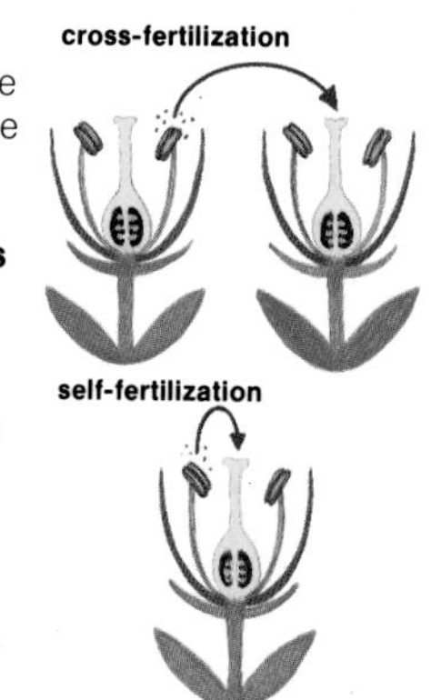

inbreeding (*n*) breeding (p. 59) over many generations (↓) between closely-related individuals (p. 135) of a species (p. 134).
outbreeding (*n*) breeding (p. 59) between individuals (p. 135) that are not closely related.
compatible (*adj*) of two plants which are able to breed (p. 59) with each other. **compatibility** (*n*).
self-compatible (*adj*) of an individual (p. 135) plant which can fertilize (↑) its female gametes (p. 61) with its own male gametes.
incompatible (*adj*) of two plants which cannot breed (p. 59) with each other. **incompatibility** (*n*).
self-incompatible (*adj*) of an individual (p. 135) plant which cannot fertilize (↑) its female gametes (p. 61) with its own male gametes.
hybrid (*n*) a plant which results from the cross-fertilization (↑) of two different species (p. 134), subspecies (p. 135), varieties (p. 135), strains (p. 135), etc. **hybridize** (*v*), **hybridization** (*n*).
heterosis (*n*) the condition of a hybrid (↑) that is fitter than either of its parents. This is also called hybrid vigour (↓).
hybrid vigour = heterosis (↑).
generation (*n*) a set of individuals (p. 135) of roughly equal age or stage of development (p. 109). The parents are one generation, and the progeny (p. 59) are the next.

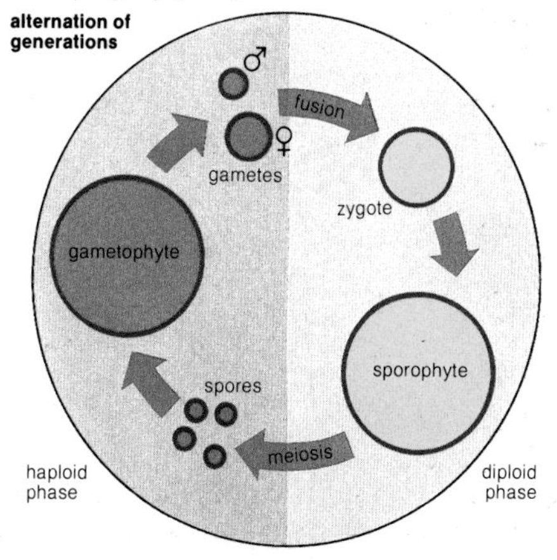

life cycle the complete set of changes that occur from any stage in the life of an organism to the same stage in the life of its offspring (p. 44). In bryophytes (p. 122), pteridophytes (p. 126) and spermatophytes (p. 128), the life cycle consists of an alternation of haploid (p. 50) and diploid (p. 50) generations (p. 63).

alternation of generations the life cycle (↑) of bryophytes (p. 122), pteridophytes (p. 126), and spermatophytes (p. 128), which consists of a haploid (p. 50) gametophyte (↓) producing gametes (p. 61) followed by a diploid (p. 50) sporophyte (↓) producing spores (p. 66).

haplont (*adj*) of the haploid (p. 50) stage in a life cycle (↑), ending with fertilization (p. 62), e.g. of the gametophyte (↓).

diplont (*adj*) of the diploid (p. 50) stage in a life cycle (↑), e.g. of the sporophyte (↓).

alternation of generations and the major plant divisions

	gametophyte haploid	**sporophyte** diploid	
bryophytes	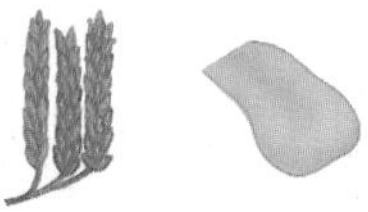	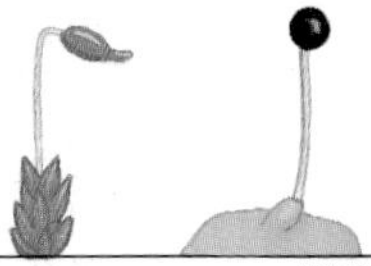	sporophyte dependent on gametophyte
pteridophytes	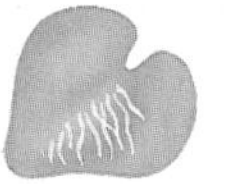	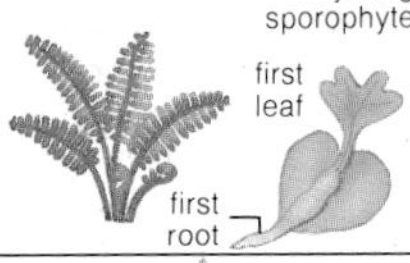 young sporophyte; first leaf; first root	sporophyte dependent on gametophyte only in very young stage
gymnosperms	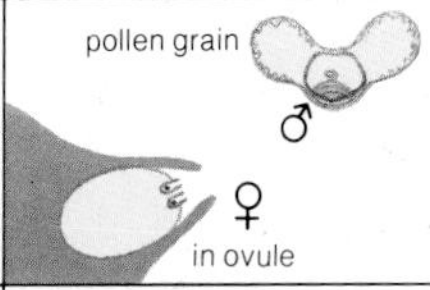 pollen grain ♂; ♀ in ovule		gametophyte dependent on sporophyte
angiosperms	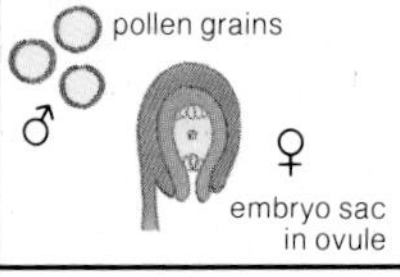 pollen grains ♂; ♀ embryo sac in ovule		gametophyte dependent on sporophyte

archegonia

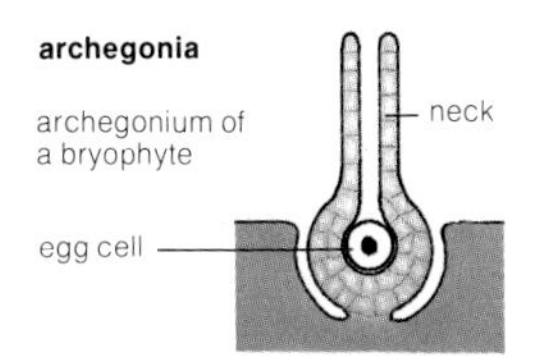

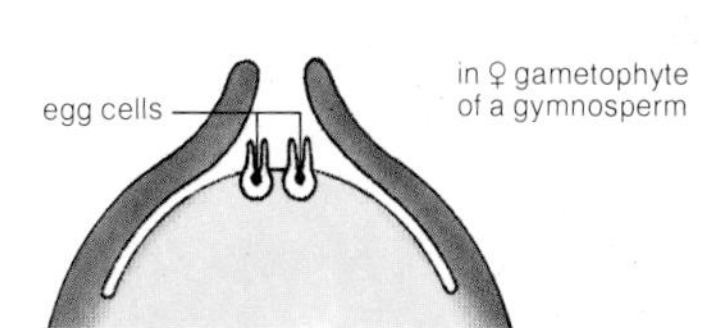

position of archegonia and antheridia in a thalloid liverwort

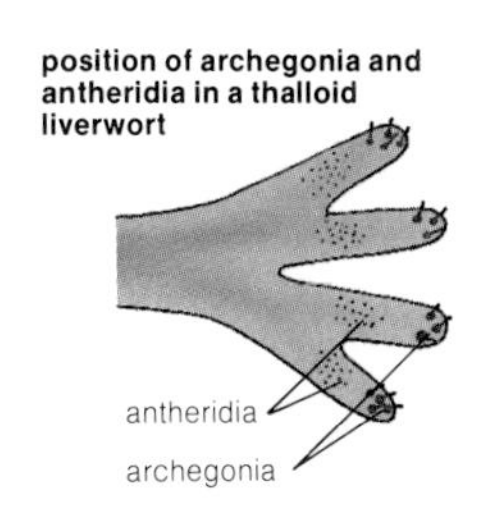

sexual reproduction in a bryophyte

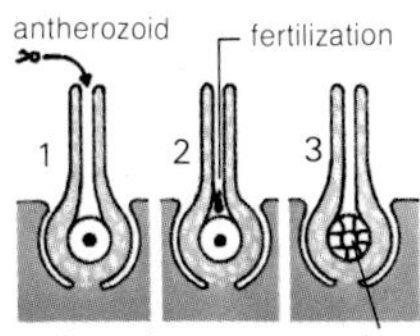

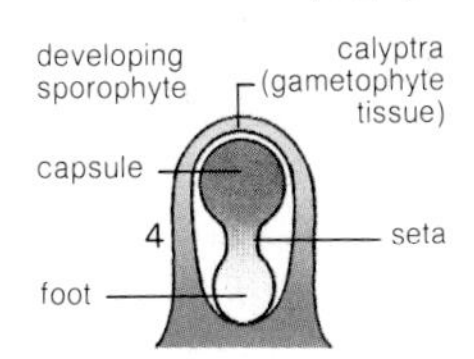

cross section of antheridium of a liverwort

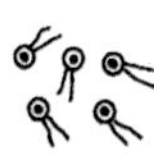

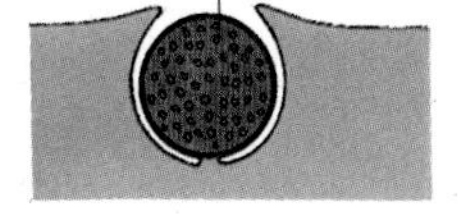

gametophyte (*n*) the haploid (p. 50) generation (p. 63) in an alternation of generations (↑). The gametophyte is the generation producing gametes (p. 61). In bryophytes (p. 122), the gametophyte is the main vegetative (p. 60) stage. In angiosperms (p. 130), the gametophyte is very small, contained in the ovules (p. 78) and pollen (p. 74) grains.

sporophyte (*n*) the diploid (p. 50) generation (p. 63) in an alternation of generations (↑). The sporophyte is the generation producing spores (p. 66). In angiosperms (p. 130), gymnosperms (p. 128) and pteridophytes (p. 126), the sporophyte is the main vegetative (p. 60) stage. In bryophytes (p. 122), the sporophyte grows directly from the archegonium (↓) of the gametophyte (↑), and depends on the gametophyte for its nutrition (p. 111).

gametangium (*n*) any organ (p. 88) which produces gametes (p. 61). **gametangia** (*pl.*).

archegonium (*n*) the flask-shaped female organ (p. 88) of bryophytes (p. 122), pteridophytes (p. 126) and gymnosperms (p. 128). The archegonium consists of a hollow neck, whose wall is one cell thick, and a swollen base containing the ovum (p. 61). The antherozoid (↓) swims down the neck to reach the ovum. **archegonia** (*pl.*), **archegoniate** (*adj*).

antheridium (*n*) the organ producing male gametes (p. 61) in bryophytes (p. 122) and ferns (p. 126). **antheridia** (*pl.*).

antherozoid (*n*) a flagellate (p. 121), motile (p. 121) male gamete (p. 61) of bryophytes (p. 122) and some ferns (p. 126). Antherozoids are produced in antheridia (↑).

spermatozoid (*n*) a motile (p. 121) male gamete (p. 61), or antherozoid (↑), in bryophytes (p. 122), ferns (p. 126) and many algae (p. 119).

spore (*n*) a small round cell with a thick wall from which a whole new plant is produced. In bryophytes (p. 122), pteridophytes (p. 126) and spermatophytes (p. 128), spores are haploid (p. 50) and are produced by the sporophyte (p. 65). In bryophytes and pteridophytes, dispersal (p. 84) is achieved by spores. In angiosperms (p. 130), the spores develop (p. 109) into small gametophytes (p. 65) in the ovules (p. 78) and pollen (p. 74) grains. In all these plants, spores are produced as a result of meiosis (p. 49). Fungi (p. 163) also produce spores, but these are of many kinds and are different from those of green plants (*see* p. 163).
spore mother cell a cell which divides by meiosis (p. 49) to produce spores (↑).
tetrad[2] (*n*) a group of four haploid (p. 50) spores (↑), which are the product of meiosis (p. 49) of the spore mother cell (↑).
sporogenous (*adj*) of tissues (p. 88) in which spores (↑) are produced.
sporulation (*n*) the process of releasing spores (↑) for dispersal (p. 84). **sporulate** (*v*).

production of haploid spores in sporangia of vascular plants

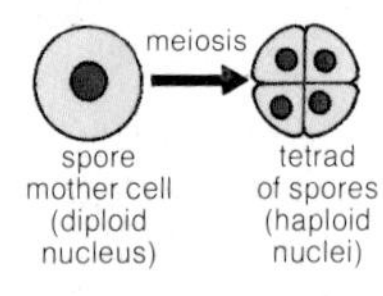

sporangia

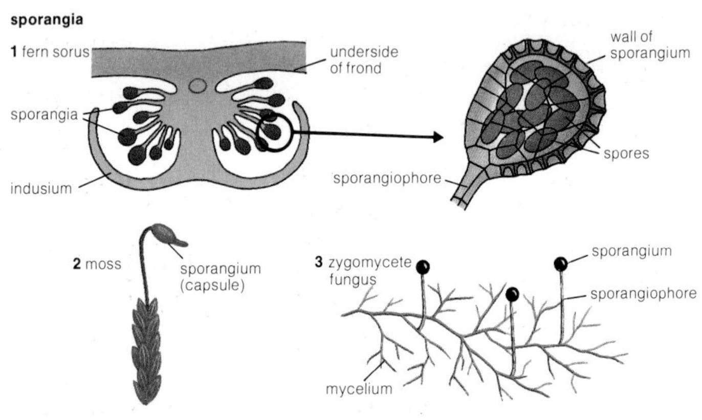

sporangium (*n*) a small round organ (p. 88) in which spores (↑) are produced, by meiosis (p. 49), from spore mother cells (↑). **sporangia** (*pl.*).
sporangiophore (*n*) the stalk of a sporangium (↑).

sporophyll (*n*) a modified leaf, whose function is to produce sporangia (↑) and spores (↑). Sporophylls may be similar to vegetative (p. 60) leaves, as in many pteridophytes (p. 126), or organized into cones (p. 68), as in gymnosperms (p. 128). The sporophylls of angiosperms (p. 130) are the stamens (p. 73) and carpels (p. 75).

homosporous (*adj*) of plants whose spores (↑) are all the same, as in bryophytes (p. 122) and true ferns (p. 126). **homospory** (*n*).

homospory and heterospory in vascular plants

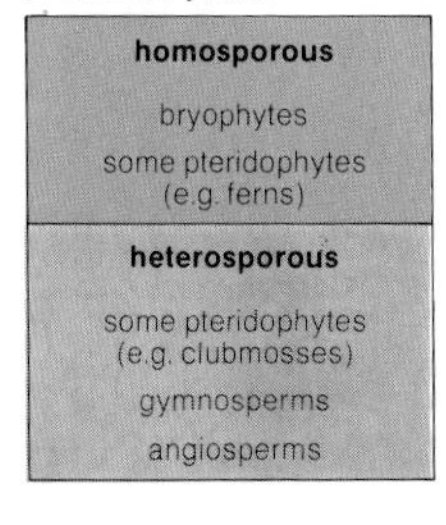

homosporous
bryophytes
some pteridophytes (e.g. ferns)
heterosporous
some pteridophytes (e.g. clubmosses)
gymnosperms
angiosperms

heterosporous (*adj*) of plants which produce spores (↑) of two different sizes, as in some pteridophytes (p. 126) and all spermatophytes (p. 128). The large spore develops into a female gametophyte (p. 65), and the small spore develops into a male gametophyte. **heterospory** (*n*).

microspore (*n*) a small spore (↑), produced in a microsporangium (↓), in heterosporous (↑) plants. The microspore develops into the male gametophyte (p. 65). In angiosperms (p. 130), the microspore is the pollen (p. 74) grain.

heterosporous plants the production of microspores

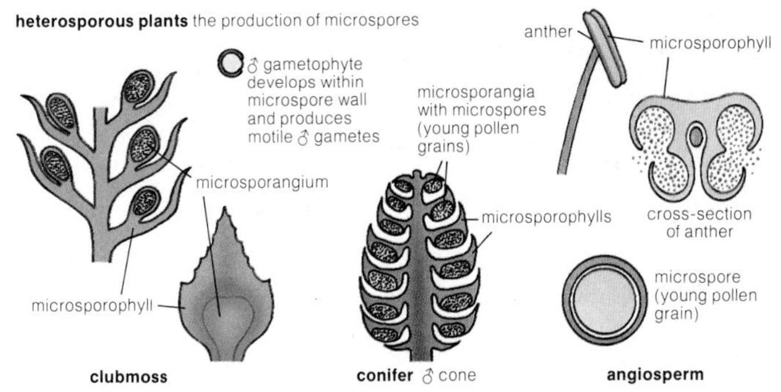

microsporangium (*n*) a sporangium (↑) producing microspores (↑) in a heterosporous (↑) plant. Microsporangia (*pl.*) usually produce many more spores than megasporangia (p. 68).

microsporophyll (*n*) a sporophyll (↑) bearing microsporangia (↑).

heterosporous plants the production of megaspores

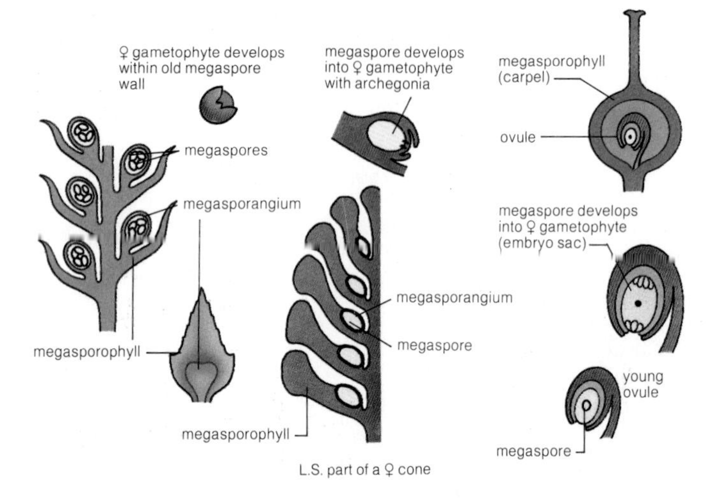

megaspore (*n*) a large spore (p. 66) produced in a megasporangium (↓), in heterosporous (p. 67) plants. The megaspore develops into the female gametophyte (p. 65). In angiosperms (p. 130), the megaspore is the embryo sac (p. 78).

megasporangium (*n*) a sporangium (p. 66) producing megaspores (↑), in heterosporous (p. 67) plants. **megasporangia** (*pl.*).

megasporophyll (*n*) a sporophyll (p. 67) bearing megasporangia (↑). In angiosperms (p. 130), the carpels (p. 75) are the megasporophylls.

cone (*n*) a group of sporophylls (p. 67) closely packed together around a central axis (p. 92). Cones are the reproductive (p. 59) structures of all gymnosperms (p. 128) and many pteridophytes (p. 126). In many cone-bearing plants, the male and female cones are separate.

strobilus (*n*) a reproductive (p. 59) organ (p. 88) consisting of overlapping scales (p. 100), as in some pteridophytes (p. 126) and the cones (↑) of gymnosperms (p. 128). **strobili** (*pl.*).

propagation (*n*) the process of reproduction (p. 59), either by natural or artificial means. **propagate** (*v*).

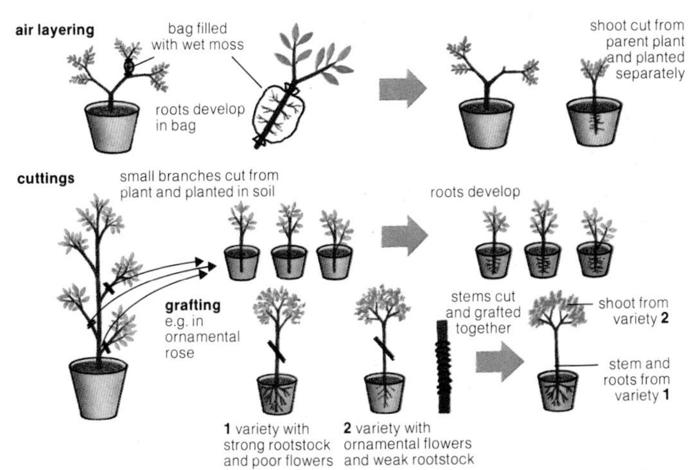

air layering a way of causing the formation of roots from nodes (p. 90) on a shoot. Wet moss (p. 124) is wrapped around the shoot; when roots have formed, the shoot can be cut from the plant and be grown as a new individual (p. 135).

cutting (*n*) a piece of shoot cut from a plant, which grows roots from its nodes (p. 90) when placed in soil.

graft (*v*) to join together artificially parts from two different plants, e.g. the shoot of one variety (p. 135) of a species (p. 134) onto the rootstock (↓) of another variety. **graft** (*n*).

rootstock (*n*) the roots of a plant.

tissue culture a process in which cells from an organism are grown on a medium (p. 171) in isolation from the organism from which they were taken. Plant tissue cultures, which usually consist of calluses (↓) of undifferentiated (p. 110) cells, are sometimes used for the production of drugs.

callus[1] (*n*) a lump of undifferentiated (p. 110) cells in a tissue culture (↑).

flower (*n*) the reproductive (p. 59) shoot of an angiosperm (p. 130), consisting usually of four sets of modified leaves arranged in whorls (p. 98). These are the sepals (↓), petals (↓), stamens (p. 73) and carpels (p. 75). The function of a flower is to produce male gametes (p. 61), in pollen (p. 74), and female gametes, in ovules (p. 78). After fertilization (p. 62), the ovules develop into seeds. The reproductive shoots of conifers (p. 128) are also sometimes called flowers. **floral** (*adj*).

perianth

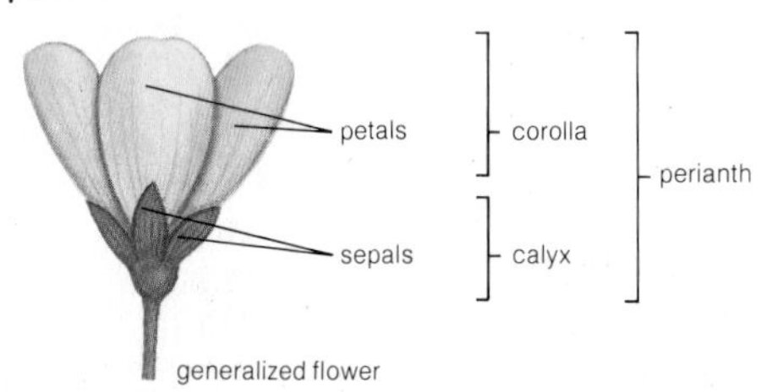

generalized flower

perianth (*n*) the outer whorls (p. 98) of the flower (↑), i.e. the calyx (↓) and corolla (↓), which are the parts of the flower not concerned with the production of gametes (p. 61). The function of the perianth is to protect the reproductive (p. 59) organs (p. 88), and to attract pollinators (p. 74) to the flower.

calyx (*n*) the outer whorl (p. 98) of the perianth (↑), consisting of sepals (↓).

sepal (*n*) a usually green, leaf-like organ (p. 88). A whorl (p. 98) of sepals forms the calyx (↑) of a flower (↑). The sepals are the outer layer of the flower bud (p. 110) before it opens.

corolla (*n*) the inner whorl (p. 98) of the perianth (↑) of a flower (↑), composed of petals (↓).

petal (*n*) an often brightly coloured leaf-like organ (p. 88). A whorl (p. 98) of petals forms the corolla (↑) of a flower (↑). The function of coloured petals is often to attract pollinators (p. 74) to the flower.

tepal (*n*) an organ (p. 88) of a perianth (↑) in which there is no difference between the calyx (↑) and corolla (↑), e.g. in tulips.

L.S. of flower

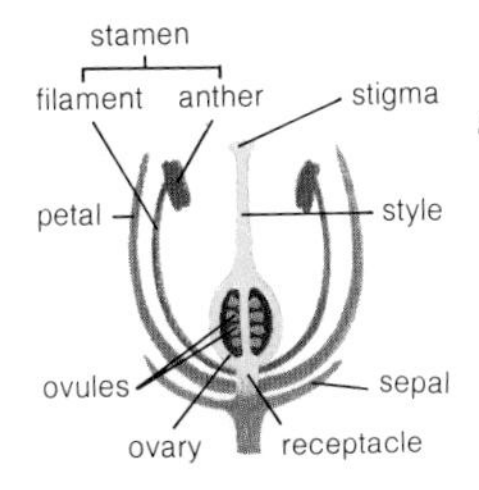

floral diagram
a flower with 6 petals, 6 stamens, 6 sepals

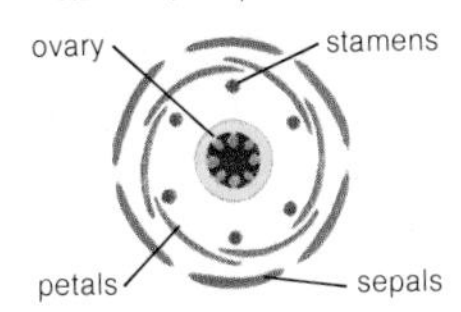

floral diagram a diagram showing the position and number of all the parts of a flower (↑) in transverse section (p. 171).

actinomorphic (*adj*) of flowers (↑) which are symmetrical (↓) in all directions (radially symmetrical) when viewed from above, that is, with each whorl (p. 98) consisting of organs (p. 88) of the same size.

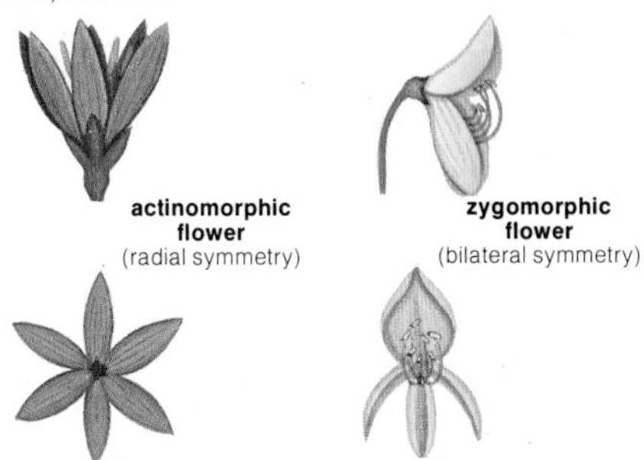

zygomorphic (*adj*) of flowers (↑) which are symmetrical (↓) in one direction only (bilaterally symmetrical), e.g. the flowers of orchids, often due to differences in sizes and shapes of petals (↑) and/or sepals (↑).

symmetrical (*adj*) of structures whose parts are arranged equally and regularly on either side of a line or plane (bilaterally symmetrical), e.g. in a zygomorphic (↑) flower, or around a central point (radially symmetrical), e.g. in an actinomorphic (↑) flower. **symmetry** (*n*).

asymmetrical (*adj.*) not symmetrical (↑).

apetalous (*adj*) of flowers (↑) without petals (↑). Apetalous flowers are often pollinated (p. 74) by wind.

gamopetalous (*adj*) of flowers (↑) in which the corolla (↑) is a tube.

polypetalous (*adj*) of flowers (↑) with petals (↑) that are not united.

sympetalous (*adj*) = gamopetalous (↑).

gamosepalous (*adj*) of flowers (↑) in which the sepals (↑) are united at their margins (p. 97).

polysepalous (*adj*) of flowers (↑) with sepals (↑) that are not united.

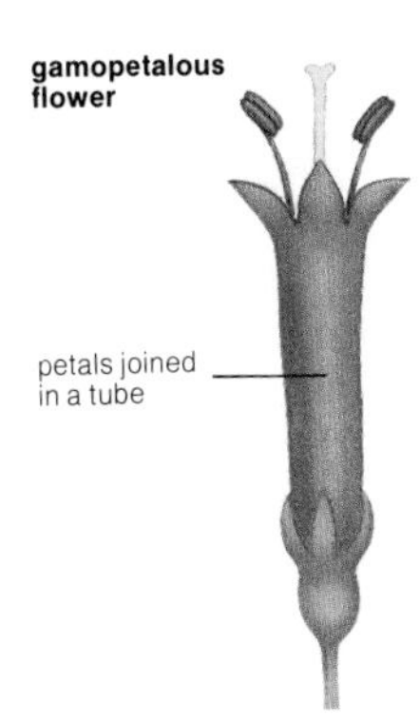

receptacle (*n*) the top of the stalk of a flower, bearing the perianth (p. 70), stamens (↓) and pistil (p. 75).

torus (*n*) the name sometimes given to the receptacle (↑) of a flower.

disk (*n*) a flat, circular receptacle (↑).

aestivation (*n*) the way in which parts of the flower, i.e. calyx (p. 70), corolla (p. 70), stamens (↓) and pistil (p. 75) are arranged with respect to each other.

hypogynous (*adj*) of flowers in which the stamens (↓), petals (p. 70) and sepals (p. 70) grow from below the gynoecium (p. 75) on the receptacle (↑). **hypogyny** (*n*).

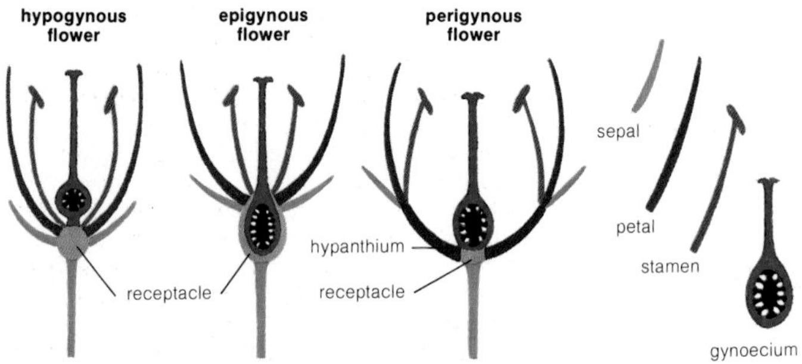

epigynous (*adj*) of flowers in which the ovary (p. 76) is within the receptacle (↑), and the other floral parts attached above it. **epigyny** (*n*).

hypanthium (*n*) a tube which results from growth at the edge of the receptacle (↑), in some plants. The perianth (p. 70) and the stamens (↓) grow from the top of the hypanthium.

perigynous (*adj*) of flowers with a hypanthium (↑). **perigyny** (*n*).

nectary (*n*) a gland (p. 112) which secretes (p. 112) nectar (↓). Many angiosperms (p. 130) have nectaries in their flowers; animals feed on the nectar and at the same time carry pollen (p. 74) from one flower to another. Some plants have extrafloral (↓) nectaries, providing food for ants which protect the plant against herbivores (p. 153).

male floral parts

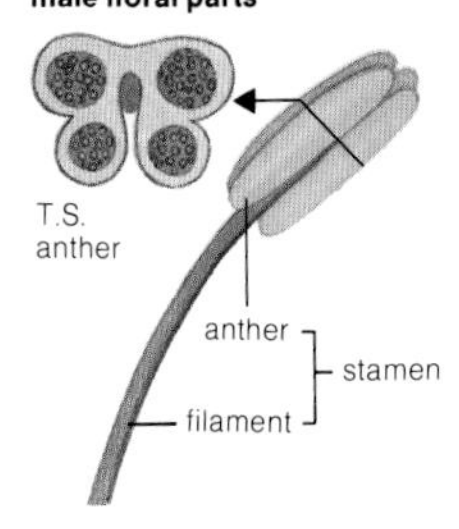

nectar (*n*) a liquid containing sugars, amino acids (p. 56), and other organic (p. 11) compounds. Nectar is secreted (p. 112) by nectaries (↑).

extrafloral (*adj*) positioned away from the flower, e.g. an extrafloral nectary (↑).

anthesis (*n*) flower-opening.

androecium (*n*) the male part of a flower, consisting of stamens (↓). The function of the androecium is to produce the male gametes (p. 61) contained in pollen (p. 74).

stamen (*n*) the male reproductive (p. 59) organ (p. 88) of a flower, consisting of a filament (↓) bearing an anther (↓). The stamen is attached to the receptacle (↑) between the petals (p. 70) and the pistil (p. 75). The number, shape, and position of the stamens in a flower are important characters in the classification (p. 132) of angiosperms (p. 130). **staminal** (*adj*).

staminate (*adj*) of flowers which have stamens (↑) but no pistil (p. 75), i.e. male flowers.

staminode (*n*) a sterile (p. 62) stamen (↑), which does not produce pollen (p. 74).

anther (*n*) the part of a stamen (↑) in which pollen (p. 74) is produced. The anther is attached to the receptacle by the filament (↓). Anthers are hollow organs (p. 88) which dehisce (p. 84) along one side to release pollen.

filament (*n*) the stalk of a stamen (↑). The filament attaches the anther (↑) to the receptacle (↑) of the flower.

basifixed (*adj*) of an organ (p. 88) which is attached to another organ by its base. This is one way in which anthers (↑) are attached to filaments (↑).

monadelphous (*adj*) having all stamens (↑) joined together in a tube which surrounds the style (p. 76), e.g. lupin.

diadelphous (*adj*) having stamens (↑) in two groups, with the stamens in each group joined together by their filaments (↑), e.g. pea.

polyadelphous (*adj*) of stamens (↑) which are united by their filaments (↑) into three or more groups in a flower.

dimorphic (*adj*) having two shapes, e.g. two different kinds of stamen (↑) in one flower.

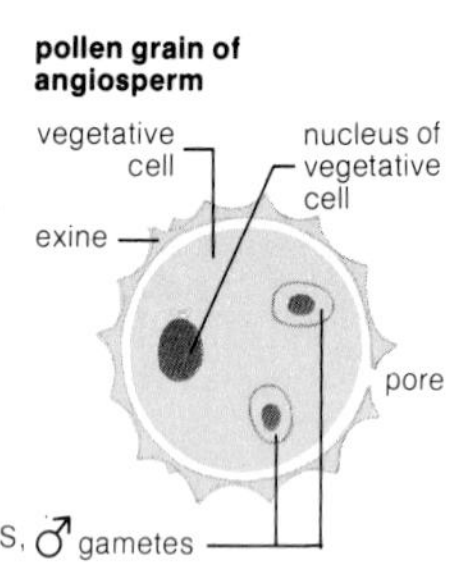

pollen (*n*) small grains containing the male gametophyte (p. 65), in seed plants (p. 128). A pollen grain has a hard wall, or exine (↓). The gametophyte consists of just three cells in angiosperms (p. 130), and between four and forty cells in gymnosperms (p. 128); in both cases, only two cells are gametes (p. 61). The pollen grain protects the male gametophyte during its journey to the female reproductive (p. 59) organs (p. 88). In angiosperms, pollen is produced in the anthers (p. 73); in gymnosperms, it is produced in male cones (p. 68).

exine (*n*) the hard outer coat of a pollen (↑) grain. The patterns of the surface of the exine are often used as characters in the classification (p. 132) of seed plants (p. 128).

sporopollenin (*n*) the material contained in the exine (↑) of pollen (↑) grains. Sporopollenin is resistant to decay and under the right conditions the exine may last for many thousands of years, although its contents die.

pollen sac the hollow space inside an anther (p. 73), where pollen (↑) grains are produced.

pollen tube a thread of cytoplasm (p. 18), covered by a membrane (p. 18), which grows from the pollen (↑) grain into the micropyle (p. 85) of the ovule (p. 78), through the tissues (p. 88) of the style (p. 76). In angiosperms (p. 130), the pollen tube carries two haploid (p. 50) nuclei (p. 19) into the ovule, one to fertilize (p. 62) the ovum (p. 61), and the other to fertilize the endosperm (p. 86) nucleus. The pollen tube will only grow if the pollen grain has landed on the stigma (p. 76).

pollination (*n*) the process in which pollen (↑) is carried from the anther (p. 73) to the surface of the stigma (p. 76), in angiosperms (p. 130), or from the male cone (p. 68) to the female cone in gymnosperms (p. 128). This can happen in several ways, e.g. by wind, water, insects, birds, bats, or even non-flying mammals, depending on the species (p. 134) of the plant. **pollinate** (*v*), **pollinator** (*n*).

cross-pollination (*n*) pollination (↑) of one plant by pollen (↑) from another individual (p. 135).

self-pollination (*n*) the pollination (↑) of an ovule (p. 78) by pollen (↑) from the same flower or the same individual (p. 135).

vector[1] (*n*) anything which carries pollen (↑) from one plant to another, e.g. insects, birds, wind, etc.

entomophily (*n*) pollination (↑) by insects. Flowers pollinated by insects are usually brightly coloured and scented. If they are pollinated by bees, they usually produce large amounts of pollen (↑) which the bees collect. If they are pollinated by butterflies or moths, they produce nectar (p. 73). **entomophilous** (*adj*).

honey guides coloured spots or lines on the petals (p. 70) of a flower, which may guide pollinating animals towards the sources of pollen (↑) and nectar (p. 73).

ornithophily (*n*) pollination (↑) by birds. Ornithophilous flowers are usually brightly coloured and secrete (p. 112) nectar (p. 73) on which the birds feed. **ornithophilous** (*adj*).

anemophily (*n*) pollination (↑) by wind. Plants which are pollinated by wind produce large amounts of pollen (↑). They are not usually scented, do not produce nectar (p. 73), and are sometimes apetalous (p. 71). **anemophilous** (*adj*).

pollinium (*n*) a large group of pollen (↑) grains, which are carried together during pollination (↑), as in the orchid family (p. 134) Orchidaceae. **pollinia** (*pl.*).

gynoecium (*n*) the female part of a flower, consisting of one or more pistils (↓).

pistil (*n*) the female reproductive (p. 59) organ (p. 88) of a flower, consisting of the ovary (p. 76), style (p. 76) and stigma (p. 76).

pistillate (*adj*) of flowers which have pistils (↑) but no stamens (p. 73), i.e. female flowers.

carpel (*n*) the female reproductive (p. 59) unit of a flower, consisting of the ovary (p. 76) with ovules (p. 78). The carpels are the sporophylls (p. 67) of angiosperms (p. 130), and are like highly modified leaves. Many angiosperms have several carpels, which are joined together at their margins (p. 97) to form the ovary.

pistil

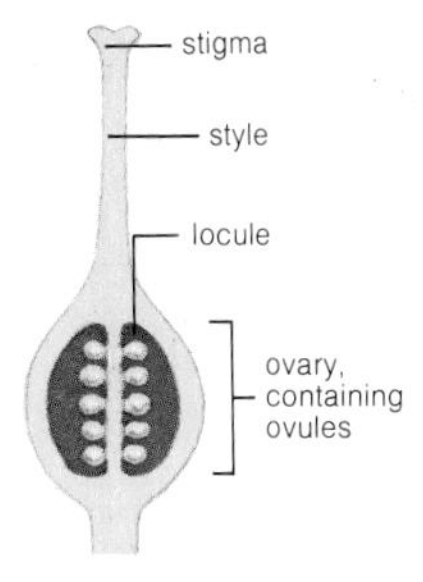

style (*n*) a long upgrowth at the top of a carpel (p. 75), bearing the stigma (↓) at its tip. The style positions the stigma so that it is likely to receive pollen (p. 74). After pollen has reached the stigma, pollen tubes (p. 74) grow down through the style to the ovary (↓).

stigma (*n*) the tip of the style (↑) of a flower. Pollen (p. 74) must reach the stigma if successful pollination (p. 74) is to occur. **stigmatic** (*adj*).

homostylous (*adj*) of plant species (p. 134) having styles (↑) all of the same length on all individuals (p. 135). **homostyly** (*n*).

heterostylous (*adj*) of plant species (p. 134) having two or more different lengths of style (↑) on different individuals (p. 135). **heterostyly** (*n*).

ovary (*n*) the interior of the carpel (p. 75) of a flower, containing the ovules (p. 78). The ovary has a thick wall, which develops into the fruit after the ovules inside it have been fertilized (p. 62) by the male gametes (p. 61) carried in the pollen (p. 74) grains.

locule (*n*) the space inside an ovary (↑).

syncarpous (*adj*) of ovaries (↑) formed by two or more carpels (p. 75) joined together. This is an important character in the classification (p. 132) of angiosperms (p. 130).

apocarpous (*adj*) of ovaries (↑) in separate carpels (p. 75), not joined together at their margins (p. 97). This is characteristic of many primitive (p. 141) flowers.

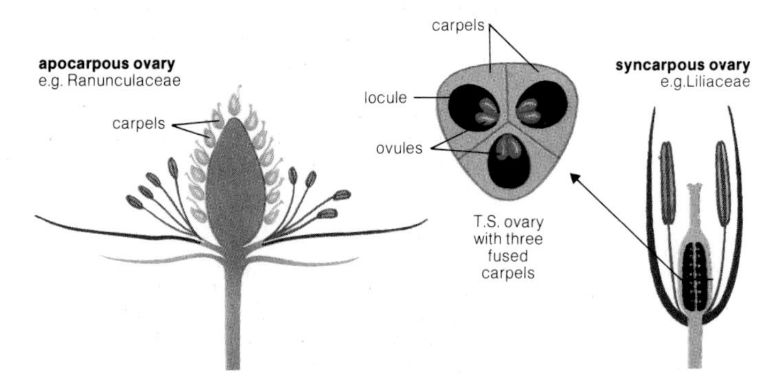

superior ovary

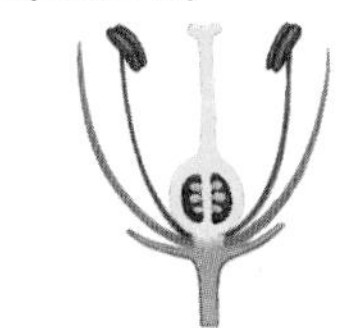

perianth and stamens attached to receptacle below ovary

inferior ovary

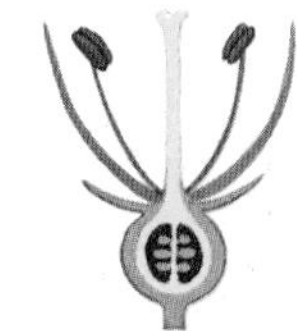

perianth and stamens attached to receptacle above ovary

inferior ovary one which is beneath the point of attachment of the calyx (p. 70), corolla (p. 70) and stamens (p. 73) of the flower.

superior ovary one which is attached to the receptacle (p. 72) above the stamens (p. 73) and the perianth (p. 70).

placentation (*n*) the arrangement of ovules (p. 78) in the ovary (↑). Because the ovules are attached to the margins (p. 97) of the carpels (p. 75), placentation depends on the way in which the carpels are joined together. Common types of placentation are axile (↓), parietal (↓) and free central (↓). This is an important character in the classification (p. 132) of angiosperms (p. 130).

placentation types
ovaries cut through to show internal structure

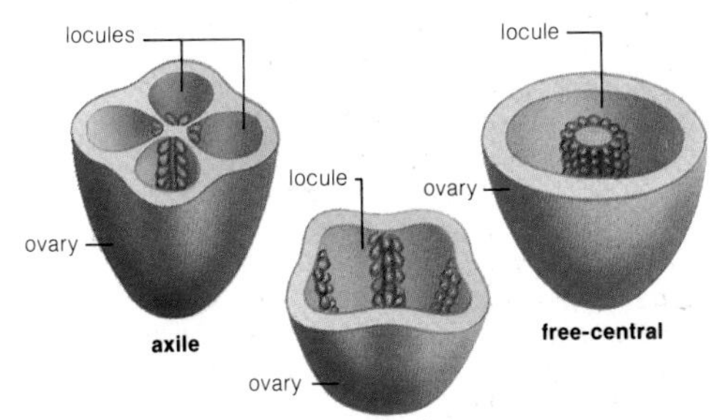

placenta (*n*) the margin (p. 97) of a carpel (p. 75), where the ovules (p. 78) are attached.

axile (*adj*) of a kind of placentation (↑) in which the margins (p. 97) of the carpels (p. 75) grow inwards to the centre of the ovary (↑), forming several locules (↑), so that the ovules (p. 78) are arranged in a divided central column.

free central (*adj*) of a kind of placentation (↑) in which the ovules (p. 78) are borne on a central growth from the bottom of the ovary (↑).

parietal (*adj*) of a kind of placentation (↑) in which the ovules (p. 78) are arranged in rows down the wall of the ovary (↑). The rows mark the lines where the margins (p. 97) of the carpels (p. 75) are joined together.

ovule (*n*) a small body in the ovary (p. 76), which contains the female gamete (p. 61), in seed plants (p. 128). After the gamete is fertilized (p. 62) by one pollen (p. 74) nucleus, the ovule develops into a seed.

funicle (*n*) the stalk of the ovule (↑), attaching it to the wall of the ovary (p. 76). After the ovule is fertilized (p. 62), the funicle becomes the stalk of the seed.

chalaza (*n*) a tissue (p. 88) in the region where the funicle (↑) is attached to the ovule (↑).

integuments (*n.pl.*) the outermost layers of the ovule (↑), which become the coat of the seed after the ovule is fertilized (p. 62).

nucellus (*n*) tissue (p. 88) of the ovule (↑), between the integuments (↑) and the embryo sac (↓).

orthotropous (*adj*) of ovules (↑) which are borne erect on the funicle (↑), with the micropyle (p. 85) pointing away from the placenta (p. 77).

campylotropous (*adj*) of ovules (↑) with the funicle (↑) attached at one side, between the chalaza (↑) and the micropyle (p. 85).

anatropous (*adj*) of ovules (↑) with the funicle (↑) bent back on itself and the micropyle (p. 85) facing the placenta (p. 77).

embryo sac the female gametophyte (p. 65) in angiosperms (p. 130), consisting of 8 haploid (p. 50) cells including the ovum (p. 61), three antipodal cells (↓), two synergids (↓) and two endosperm (p. 86) nuclei (p. 19) which fuse (p. 61) before fertilization (p. 62). The embryo sac is contained inside the ovule (↑).

synergids (*n.pl.*) a group of two cells next to the ovum (p. 61) in the female gametophyte (p. 65) of an angiosperm (p. 130).

antipodal cells the three cells at the other end of the embryo sac (↑) from the ovum (p. 61) in the female gametophyte (p. 65) of an angiosperm (p. 130).

double fertilization fertilization (p. 62) of the ovum (p. 61) by one pollen (p. 74) nucleus (p. 19), and of the endosperm mother cell (↓) by another. This happens in all angiosperms (p. 130), but not in other plants.

ovule structure

embryo sac (♀ gametophyte)
endosperm mother cell
antipodal cells
synergids
egg cells
chalaza region
micropyle
funicle
nucellus
placenta
integuments

types of ovule

orthotropous

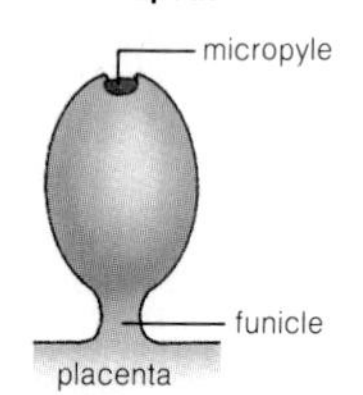

campylotropous

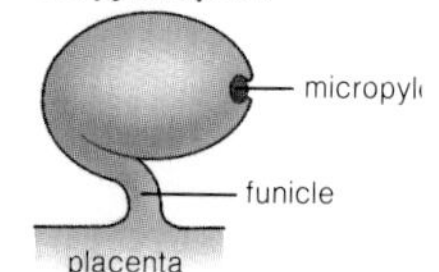

anatropous

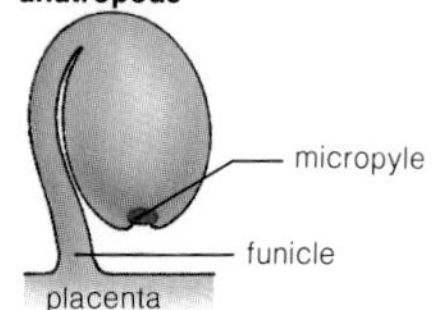

endosperm mother cell the cell formed in the embryo sac (↑) by the fusion (p. 61) of the two haploid (p. 50) endosperm (p. 86) nuclei (p. 19). The mother cell is diploid (p. 50), and it is fertilized (p. 62) by a pollen (p. 74) nucleus, in angiosperms (p. 130), to form the triploid (p. 50) endosperm.

hermaphrodite (*adj*) of flowers with male and female reproductive (p. 59) organs (p. 88).

perfect (*adj*) of flowers with male and female reproductive (p. 59) organs (p. 88), i.e. hermaphrodite (↑) flowers.

dioecious (*adj*) having male and female flowers on different individuals (p. 135) of the same plant species (p. 134). This is a way of avoiding self-fertilization (p. 62). **dioecy** (*n*).

monoecious (*adj*) having separate male and female flowers on the same individual (p. 135). **monoecy** (*n*).

gynodioecious (*adj*) having female and hermaphrodite (↑) flowers separately on different individuals (p. 135) of a plant species (p. 134). **gynodioecy** (*n*).

andromonoecious (*adj*) having male and hermaphrodite (↑) flowers on the same individual (p. 135). **andromonoecy** (*n*).

polygamous (*adj*) of plants which bear male, female and hermaphrodite (↑) flowers at the same time. **polygamy** (*n*).

homogamous (*adj*) having male and female floral parts functioning at the same time. **homogamy** (*n*).

dichogamous (*adj*) of flowers in which the male and female parts become functional at different times. This is a way of avoiding self-fertilization (p. 62). **dichogamy** (*n*).

protogynous (*adj*) of flowers in which the female parts become functional before the male parts. This is a way of avoiding self-fertilization (p. 62). **protogyny** (*n*).

protandrous (*adj*) of flowers whose anthers (p. 73) produce pollen (p. 74) before the ovules (p. 78) or stigma (p. 76) of the same flower are functional. This is a way of avoiding self-fertilization (p. 62). **protandry** (*n*).

inflorescence (*n*) a shoot bearing flowers and no leaves. An inflorescence can have one to many flowers.

peduncle (*n*) the stalk of a whole inflorescence (↑).

pedicel (*n*) the stalk of a single flower in an inflorescence (↑).

scape (*n*) a flower stalk growing from the level of the ground, as in herbaceous (p. 136) plants whose leaves form a rosette (p. 99).

raceme (*n*) a kind of inflorescence (↑) with a central axis (p. 92) bearing flowers along its length. **racemose** (*adj*).

panicle (*n*) a branched inflorescence (↑), consisting of a number of racemes (↑), as in many grasses (p. 130).

corymb (*n*) a raceme (↑) whose lower stalks are longer than the upper ones, so the inflorescence (↑) has a flat top. **corymbose** (*adj*).

cyme (*n*) a sympodial (p. 93) inflorescence (↑) growing by means of lateral (p. 92) branches, each with a flower at its apex (p. 90). **cymose** (*adj*).

umbel (*n*) an inflorescence (↑) in which all the pedicels (↑) are the same length and arise from the same point.

spike (*n*) an inflorescence (↑) with a long central axis (p. 92) and sessile (p. 100) flowers, as in many grasses (p. 130).

catkin (*n*) a spike (↑) of small, either male or female flowers, falling entire from the plant, e.g. in the willow family (p. 134), Salicaceae.

inflorescence

flower
pedicel
peduncle

inflorescence in Araceae

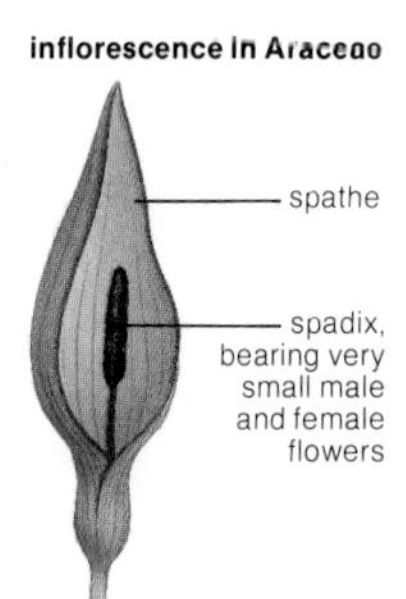

inflorescence types

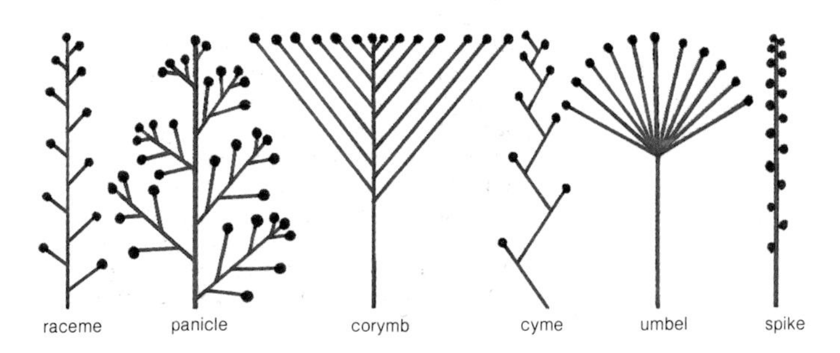

capitulum (*n*) an inflorescence (↑) like a head, consisting of many sessile (p. 100) flowers, e.g. in Compositae. **capitula** (*pl.*).

capitate (*adj*) like a head, e.g. when many flowers are clustered together in an inflorescence (↑).

composite, capitate inflorescence
many flowers in a single capitulum

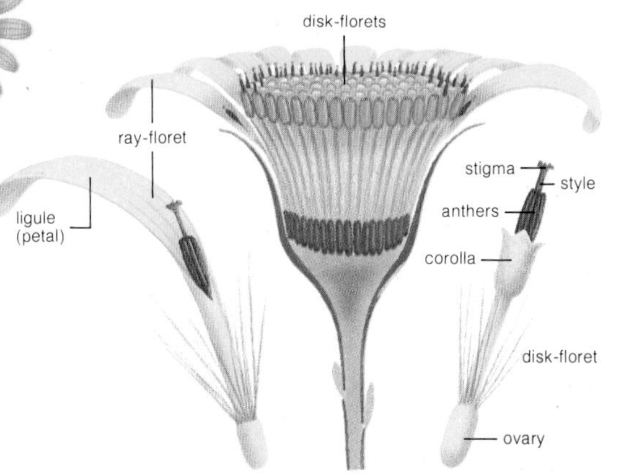

composite (*adj*) of inflorescences (↑) where many small flowers are grouped together in a head, looking like a large single flower, e.g. in the daisy family (p. 134), Compositae.

floret (*n*) a small flower, usually in a large or composite (↑) inflorescence (↑).

disk-floret a flower in the central part of a composite (↑) inflorescence (↑).

ray-floret a flower at the edge of a composite (↑) inflorescence (↑). Most ray-florets have a single petal (p. 70), called a ligule (↓).

ligule[1] (*n*) the corolla (p. 70) of a ray-floret (↑) in a composite (↑) inflorescence (↑).

involucre (*n*) a structure which protects or encloses another organ (p. 88), e.g. the bracts (p. 99) enclosing the developing inflorescence (↑) in Compositae, or the leaves joined together to protect the sex organs in leafy liverworts (p. 123).

spadix (*n*) an inflorescence (p. 80) consisting of a single fleshy (p. 99) axis (p. 92) with many small sessile (p. 100) flowers, as in the monocotyledon (p. 130) family (p. 134) Araceae. **spadices** (*pl.*).

spathe (*n*) the single large bract (p. 99) which encloses a young spadix (↑).

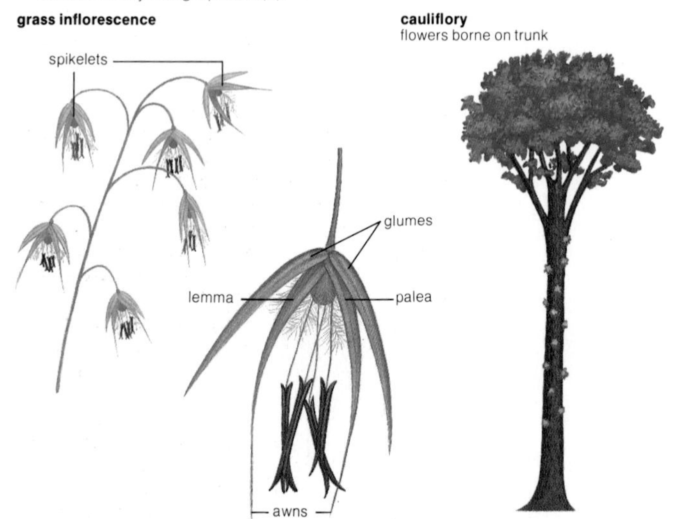

spikelet (*n*) a small branch of the spike (p. 80) in grasses (p. 130), bearing a few flowers.

lemma (*n*) one of the pair of inner bracts (p. 99) at the base of a grass (p. 130) spikelet (↑). **lemmas** (*pl.*)

glumes (*n.pl.*) the pair of outer bracts (p. 99) at the base of a grass (p. 130) spikelet (↑).

palea (*n*) one of the pair of inner bracts (p. 99) at the base of a grass (p. 130) spikelet (↑).

awn (*n*) a long, thin pointed tip, e.g. of the lemma (↑) of a grass (p. 130) flower.

cauliflorous (*adj*) of plants with flowers or inflorescences (p. 80) on the stem or trunk (p. 92). **cauliflory** (*n*).

solitary (*adj*) of organs (p. 88) which are borne singly, on their own, e.g. a flower in a one-flowered inflorescence (p. 80).

fruit (*n*) the organ (p. 88) of angiosperms (p. 130) containing the seeds. A true fruit is the product of the development of the ovary (p. 76) wall, and the seeds are fertilized (p. 62) ovules (p. 78). The function of the fruit is to protect the seeds as they develop and to help in their dispersal (p. 84). The term fruit or *fruiting body* can be used to describe any organ containing propagules (p. 59) in members of the plant kingdom (p. 134).

pome (*n*) = pseudocarp (↓).

pseudocarp (*n*) a false 'fruit' which has developed from the receptacle (p. 72), not from the ovary (p. 76), e.g. apple.

pericarp (*n*) the whole wall of the ripe (↓) ovary (p. 76) or fruit (↑), usually consisting of exocarp (↓), mesocarp (↓) and endocarp (↓).

exocarp (*n*) the outer layer of tissue (p. 88) of the fruit (↑). The exocarp is often hard or skin-like.

epicarp (*n*) = exocarp (↑).

mesocarp (*n*) the layer of tissue (p. 88) in a fruit (↑), between the exocarp (↑) and the endocarp (↑). The mesocarp is often fleshy (p. 99) or succulent (p. 99).

pulp (*n*) the succulent (p. 99) part of a fruit (↑).

endocarp (*n*) the innermost layer of tissue (p. 88) in a fruit (↑), surrounding the seeds.

ripe (*adj*) of fruits (↑) which are ready to release their seeds, or of a seed which has finished growing in the fruit. **ripen** (*v*).

monocarpic (*adj*) of plants which produce fruit (↑) only once in their life-cycle (p. 64), e.g. most annual (p. 117) plants. **monocarpy** (*n*).

parthenocarpic (*adj*) of plants whose fruits (↑) develop without seeds; this occurs naturally in some plants when fertilization (p. 62) has not taken place. **parthenocarpy** (*n*).

berry (*n*) a succulent (p. 99) or juicy fruit (↑), with many, usually small, seeds.

drupe (*n*) a fruit (↑) with seeds which are covered by a hard, stony endocarp (↑). Drupes usually have a fleshy (p. 99) mesocarp (↑).

kernel (*n*) the seed in a drupe (↑).

stone (*n*) the hard endocarp (↑) of a drupe (↑), containing the seed.

pyrene (*n*) a single stone (↑) in a small drupe (↑).

berry e.g. tomato

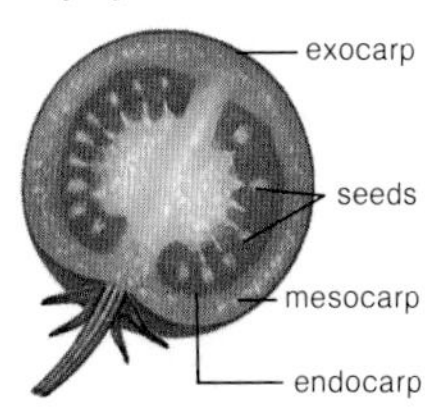

drupe e.g. apricot

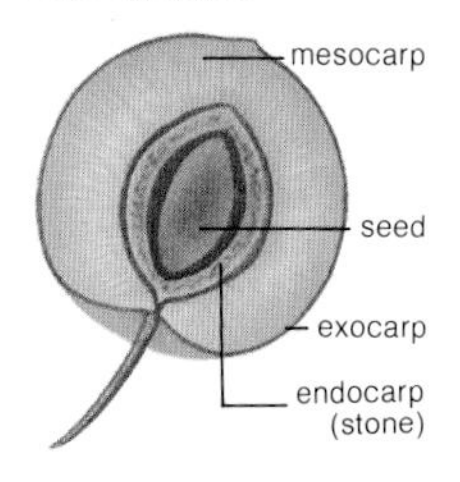

legume (*n*) a dehiscent (↓) pod (↓) containing seeds (↓), developed (p. 109) from a single carpel (p. 75). The fruit of the family (p. 134) Leguminosae (beans, clovers, acacias, etc.).

pod (*n*) a long, thin, dry fruit, developed from a single carpel (p. 75), splitting down the side where the margins (p. 97) of the carpel (p. 75) were joined.

dehisce (*v*) to split open along a line. Many fruits, especially dry fruits, dehisce to release their seeds (↓). Anthers (p. 73) dehisce to release their pollen (p. 74). **dehiscent** (*adj*).

indehiscent (*adj*) = not dehiscent (↑).

capsule[1] (*n*) a dehiscent (↑) dry fruit, with at least one carpel (p. 75), often with many small seeds (↓), e.g. in the family (p. 134) Orchidaceae.

loculicidal (*adj*) of the dehiscence (↑) of a capsule (↑) with several carpels (p. 75), which split lengthways, exposing the seeds (↓) in each locule (p. 76).

nut (*n*) a dry, indehiscent (↑) fruit with a hard wall, containing one seed (↓).

follicle (*n*) a dry, dehiscent (↑) fruit formed from a single carpel (p. 75).

achene (*n*) a dry fruit with one seed (↓), the product of one carpel (p. 75).

samara (*n*) a small dry fruit or achene (↑), with wing-like outgrowths which assist in dispersal (↓) by wind.

schizocarp (*n*) a dry fruit developed from a syncarpous (p. 76) ovary (p. 76). A schizocarp splits into achene (↑)-like units when ripe (p. 83). Each unit is a single carpel (p. 75).

silicula (*n*) a long dry fruit, developed from an ovary (p. 76) consisting of two carpels (p. 75), as in the family (p. 134) Cruciferae.

siliqua (*n*) = silicula (↑).

pappus (*n*) a group of fine hairs on a small dry fruit, which helps in dispersal (↓) by wind, e.g. in the family (p. 134) Compositae.

dispersal (*n*) the movement of propagules (p. 59) away from the parent plant, e.g. by wind or birds. Dispersal is the way in which plants can spread. Fruits and seeds have many different adaptations (p. 141) for different kinds of dispersal.

legume e.g. pea

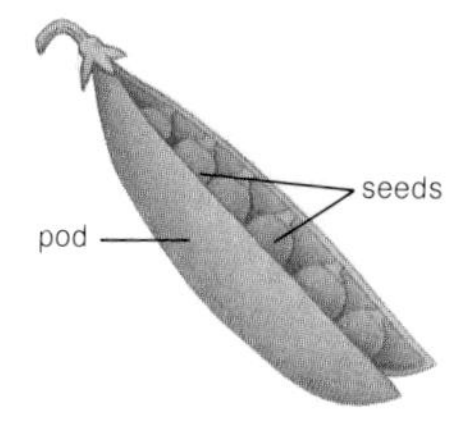

capsule e.g. poppy

achene e.g. strawberry

samara e.g. sycamore

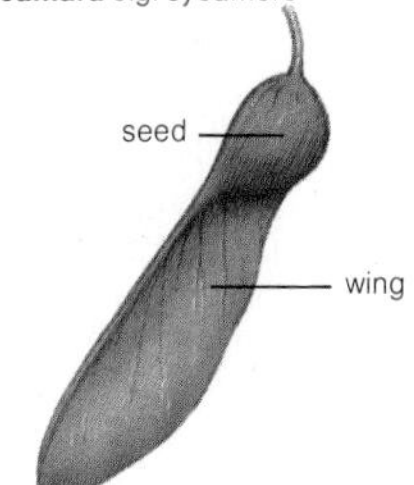

seed (*n*) a fertilized (p. 62) ripe (p. 83) ovule (p. 78) of an angiosperm (p. 130) or gymnosperm (p. 128). The seed is the product of sexual (p. 59) reproduction (p. 59), and the means by which the progeny (p. 59) of a plant can be spread. The seed is covered by a testa (↓), and contains an embryo (↓) and endosperm (p. 86). The seeds of angiosperms are produced in fruits, and those of gymnosperms are produced in cones (p. 68) or strobili (p. 68).

testa (*n*) the hard, outer coat of a seed (↑), which protects the embryo (↓) and prevents water from entering the seed until it is ready to germinate (p. 87).

hilum (*n*) the place on the seed (↑) marking the point where the funicle (p. 78) was attached to the ovule (p. 78).

micropyle (*n*) a hollow tube or pore (p. 19) at the tip of an ovule (p. 78), through which the pollen tube (p. 74) enters. The micropyle can be seen in the testa (↑) of the mature seed (↑). Water enters the micropyle at the beginning of germination (p. 87).

raphe (*n*) a long ridge on the coat of a seed (↑) which has developed from an anatropous (p. 78) ovule (p. 78). The raphe marks the position where the funicle (p. 78) of the ovule used to be.

embryo (*n*) the young plant contained in the seed (↑). The embryo is the product of repeated mitotic (p. 45) divisions of the zygote (p. 61). It consists of cotyledons (p. 86), a plumule (p. 86), a hypocotyl (p. 86) and a radicle (p. 86). **embryonic** (*adj*).

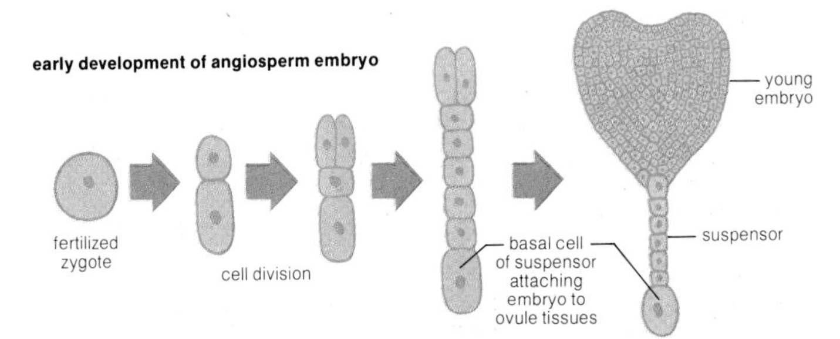

aril (*n*) an extra seed envelope, often coloured and fleshy (p. 99), found in some angiosperms (p. 130). The aril is produced from the tissues (p. 88) of the funicle (p. 78) or base of the ovule (p. 78). **arillate** (*adj*).

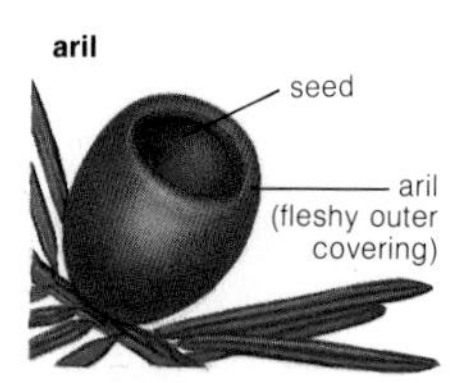

suspensor[1] (*n*) a group or chain of cells developed from the fertilized (p. 62) ovum (p. 61) in seed plants (p. 128), which attaches the embryo (p. 85) to the wall of the embryo sac (p. 78).

cotyledon (*n*) part of the embryo (p. 85) of a seed plant (p. 128). The cotyledon sometimes becomes the first photosynthetic (p. 32) organ (p. 88) of the young seedling (↓). Some plants, e.g. Leguminosae, have large cotyledons which store food. Angiosperms (p. 130) have either one or two cotyledons; gymnosperms (p. 128) have more. Angiosperms are classified (p. 132) into two classes (p. 134), the monocotyledons (p. 130) which have one cotyledon and the dicotyledons (p. 131) which have two.

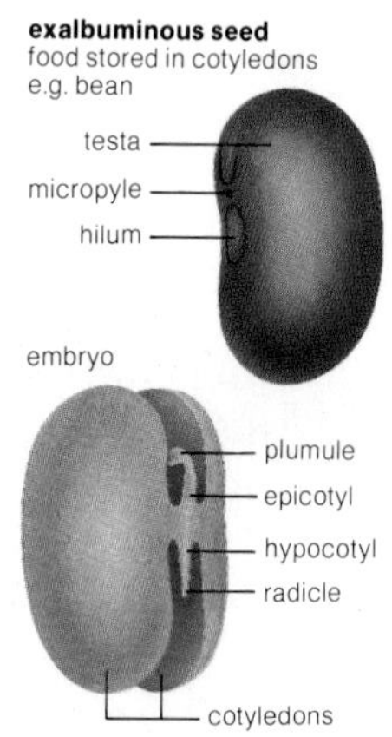

seed leaf = a cotyledon (↑).

epicotyl (*n*) part of the embryo (p. 85) and seedling (↓) above the cotyledons (↑). The first true leaves are produced on the epicotyl after germination (↓).

plumule (*n*) the apical (p. 90) part of the epicotyl (↑) of an embryo (p. 85), from which the first true leaves of the seedling (↓) develop.

hypocotyl (*n*) the part of the embryo (p. 85) and seedling (↓) below the cotyledons (↑), bearing the radicle (↓) at its end.

radicle (*n*) the part of the embryo (p. 85) that develops into the root of the seedling (↓).

endosperm (*n*) triploid (p. 50) tissue (p. 88) in the seed, resulting from double fertilization (p. 78). The function of the endosperm is to store food for the seedling (↓).

albumen (*n*) the endosperm (↑) of a seed. **albuminous** (*adj*).

exalbuminous (*adj*) of seeds without albumen (↑).

aleurone layer the outer layer of thick-walled cells in the endosperm (↑) of the seeds of many grasses (p. 130). The aleurone layer is rich in protein (p. 56).

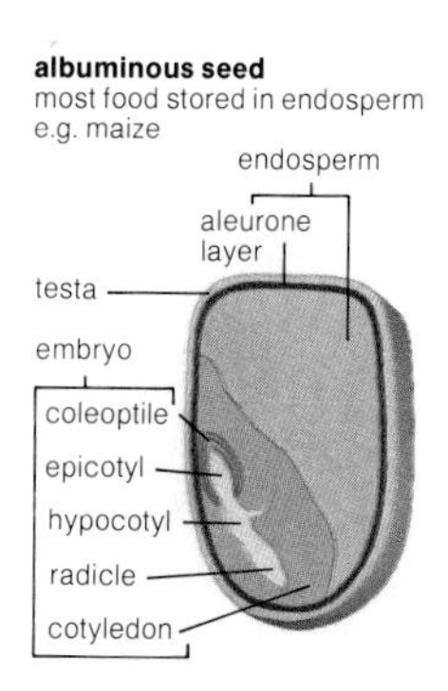

germination (*n*) the first stage in the growth of a seed into a seedling (↓), or a spore (p. 66) into a young plant. In seed plants (p. 128) germination begins with the imbibition (↓) of water and ends with the production of the first true leaves. **germinate** (*v*).

imbibition (*n*) the process in which water is taken up by a seed at the beginning of germination (↑).

epigeal (*adj*) of the kind of germination (↑) in which the cotyledons (↑) are borne above ground level, becoming the first photosynthetic (p. 32) organs (p. 88) of the seedling (↓).

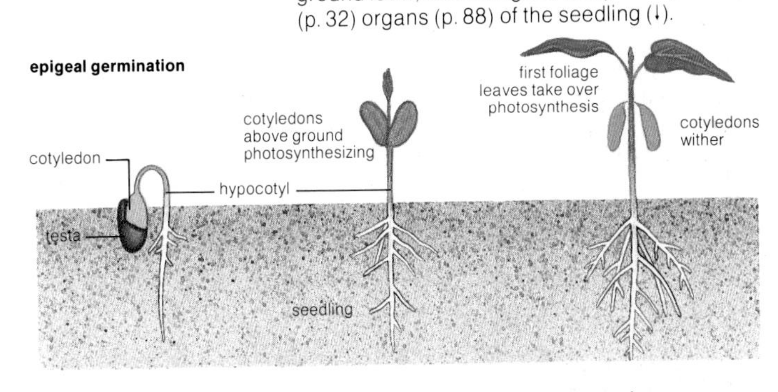

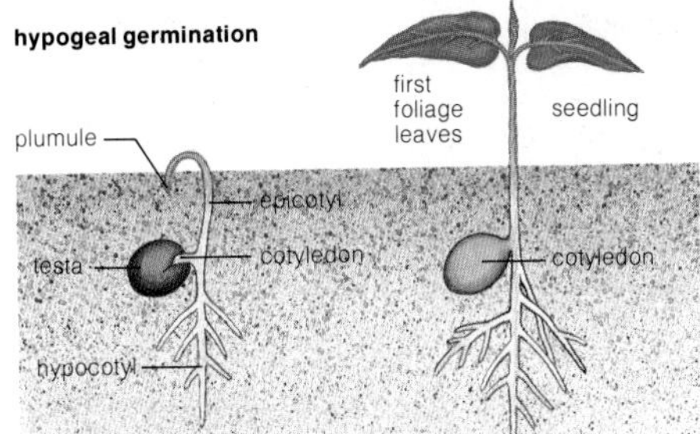

hypogeal (*adj*) of the kind of germination (↑) in which the cotyledons (↑) remain below ground. Their stored food is used up in the early growth of the epicotyl (↑) and the hypocotyl (↑).

seedling (*n*) a young plant growing from its seed. It is usually called a seedling until it loses its cotyledons (↑).

morphology (*n*) the study of the shape and arrangement of organs (↓) and tissues (↓).

anatomy (*n*) the study of the way in which tissues (↓) and organs (↓) are arranged in organisms. **anatomical** (*adj*).

tissue (*n*) a group of cells, of similar shape and size, which all have the same function. Plant organs (↓) usually have several different kinds of tissue, e.g. leaves have epidermal (p. 90), mesophyll (p. 95) and vascular (p. 122) tissue.

organ (*n*) a group of cells or tissues (↑), forming part of an organism, with a special function, e.g. a leaf, a stamen (p. 73).

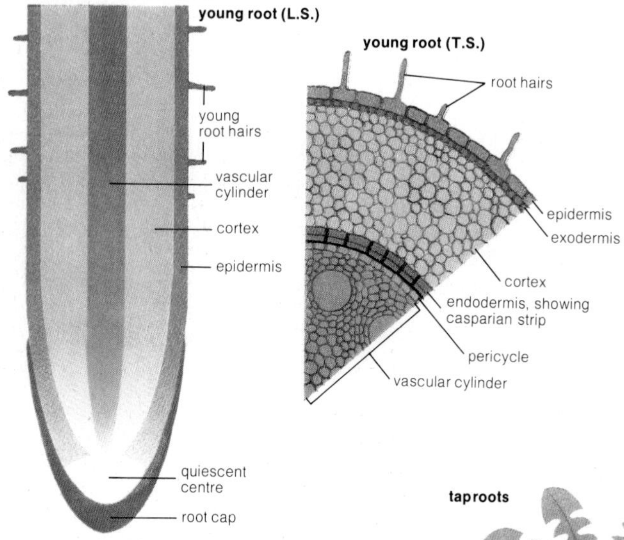

taproots

tap root

root (*n*) the organ (↑) of a plant, that grows down into the soil. Roots anchor the plant in the ground and take up water and nutrients (p. 111) from the soil. In some plants the roots also store food. They differ from stems in not having nodes (p. 90) and leaves.

radical (*adj*) of roots.

taproot (*n*) the main, primary root of a plant which shows apical dominance (p. 114).

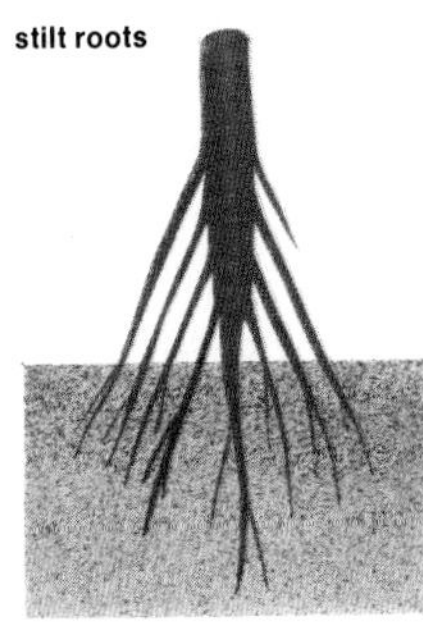

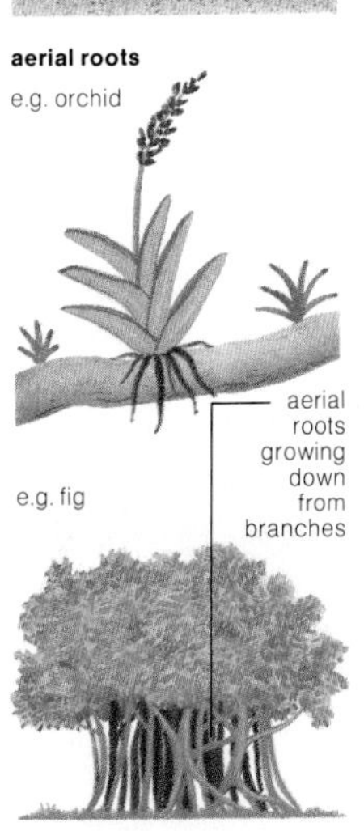

adventitious root any root which grows from a tissue (↑) other than the pericycle (↓) or endodermis (↓) of an older root.

stilt root a root which grows out from near the bottom of the trunk (p. 92), in some trees, into the ground. Its function is support. Many palms (p. 130) have stilt roots. They are sometimes known as prop roots.

prop root = stilt root (↑).

aerial root a root growing from a part of a plant which is above the ground.

velamen (*n*) the tissue (↑) of dead cells underneath the epidermis (p. 90) of the aerial roots (↑) of some plants, e.g. the orchid family (p. 134), Orchidaceae. The velamen absorbs water.

root cap a layer of cells on the surface of the root tip, which protects the root as it grows and lubricates its passage through the soil.

quiescent centre a region of cells in the root tip, at the end of the stele (p. 105), where no cell division (p. 45) takes place.

piliferous layer the layer of cells, in the epidermis (p. 90) of the root, which bears root hairs (↓).

root hair a thread-like outgrowth of a cell in the epidermis (p. 90) of a root. Root hairs increase the surface area of a root and help in the uptake (p. 101) of water and nutrients (p. 111).

endodermis (*n*) the innermost layer of the cortex (↓) of a root, surrounding the vascular cylinder (p. 105) in all vascular (p. 122) plants. **endodermal** (*adj*).

casparian strip a band of suberin (p. 94) around the cells of the endodermis (↑) of the root, which stops the movement of substances from the cortex (↓) to the vascular cylinder (p. 105) other than through the cytoplasm (p. 18) of the endodermal cells.

pericycle (*n*) a layer of cells lying inside the endodermis (↑), on the surface of the vascular cylinder (p. 105) of a root.

cortex (*n*) the tissue (↑) between the vascular cylinder (p. 105) and the epidermis (p. 90) of a root or stem. The cortex usually has many layers of cells. **cortical** (*adj*).

epidermis (*n*) the outer layer of cells of leaves, green stems, young roots, etc. **epidermal** (*adj*).

exodermis (*n*) layer of cortical (p. 89) cells, with suberin (p. 94) in their cell walls (p. 17). The exodermis is on the outer surface of the cortex, underneath the epidermis (↑). **exodermal** (*adj*).

parenchyma (*n*) the general name for tissues (p. 88) of cells with thin cell walls (p. 17), often with intercellular spaces (p. 95), e.g. the spongy mesophyll (p. 95) of leaves, or the cortex (p. 89) of stems and roots.

medulla (*n*) (1) the parenchyma (↑) or sclerenchyma (↓) inside the vascular cylinder (p. 105) of a stem or root. Its function is the storage of food. (2) the name given to the central part of the thallus (p. 122) of some algae (p. 119) and lichens (p. 147).

ray (*n*) a band of parenchyma (↑) and/or sclerenchyma (↓) cells, running from the cortex (p. 89) towards the centre of a stem.

shoot (*n*) the general name for any stem above the surface of the ground.

apex (*n*) the tip of a root or shoot. **apical** (*adj*).

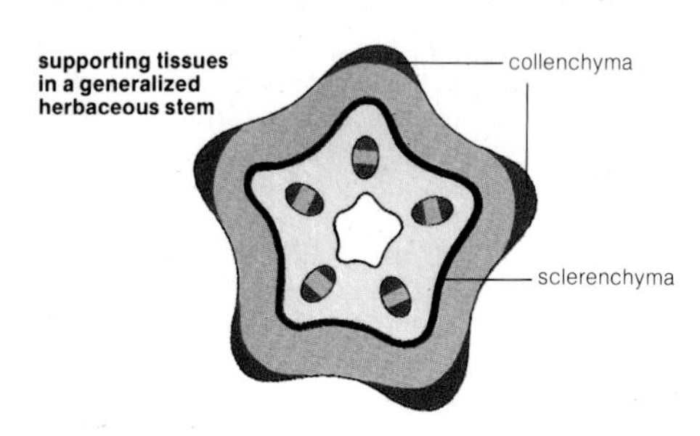

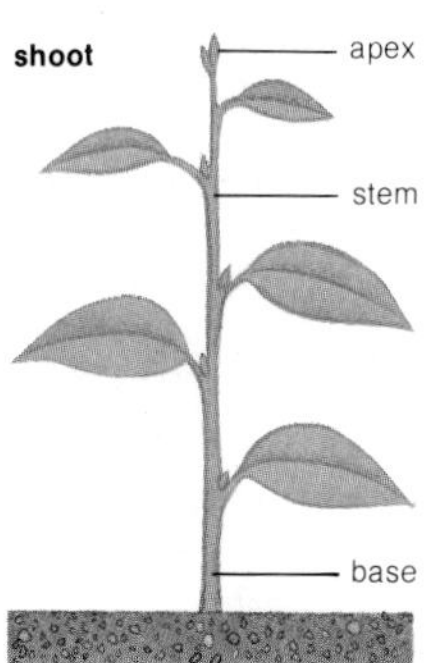

stem (*n*) the part of a plant with nodes (↓), buds (p. 110) and leaves. Most stems are above the ground, but some, e.g. rhizomes (p. 60), are underground.

node (*n*) the point on a stem from which a leaf grows. Nodes are spaced along stems, with internodes (↓) between them. **nodal** (*adj*).

internode (*n*) the space on a stem between two nodes (↑).

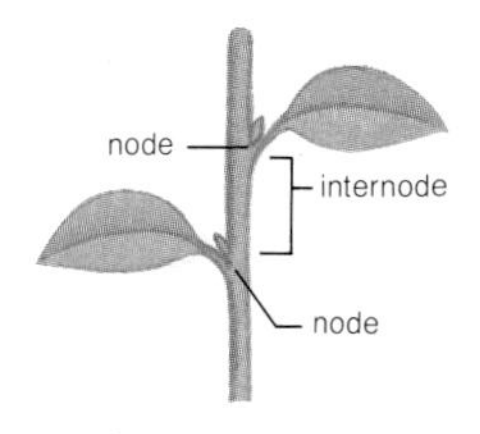

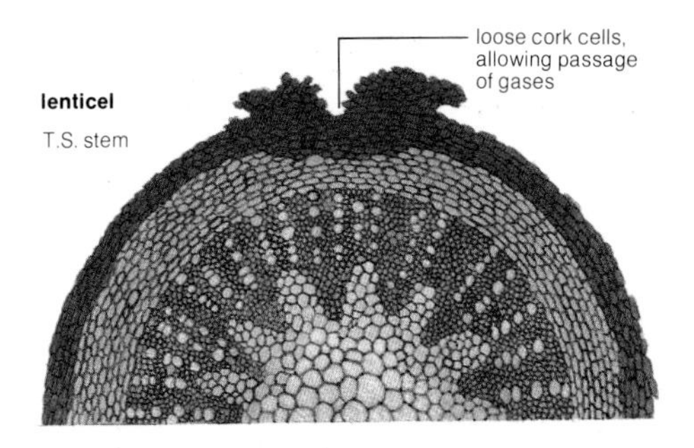

lenticel
T.S. stem

lenticel (*n*) a pore (p. 19) on the surface of the stems of some plants, allowing gas exchange between the stem and the atmosphere. **lenticellate** (*adj*).

culm (*n*) the stem of a grass.

sclerenchyma (*n*) a hard, lignified (p. 93) tissue (p. 88) consisting of fibres (↓) and sclereids (↓). It is found in the stems, roots, leaves or fruits of many plants, and its function is support.

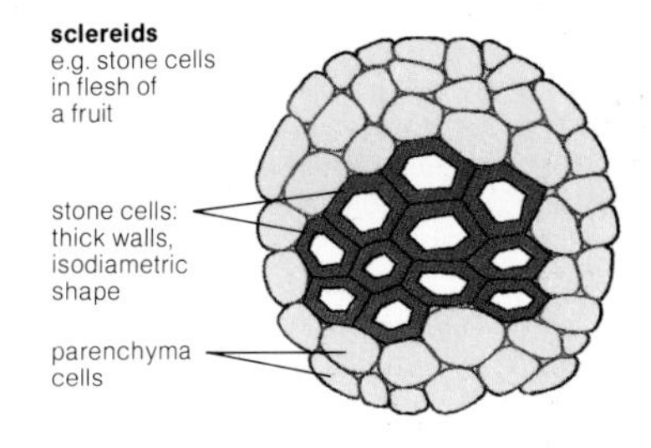

sclereids
e.g. stone cells in flesh of a fruit

sclereid (*n*) kind of cell found in the sclerenchyma (↑) of some plants, with heavily lignified (p. 93) walls. Sclereids are usually found in groups.

stone cell an isodiametric (↓) sclereid (↑).

isodiametric (*adj*) of cells or structures with sides of equal length.

fibre (*n*) a long, thick-walled cell in the sclerenchyma (↑).

sclerophyllous (*adj*) of plants whose leaves contain sclerenchyma (↑). Such leaves are usually thick and leathery.

collenchyma in a leaf

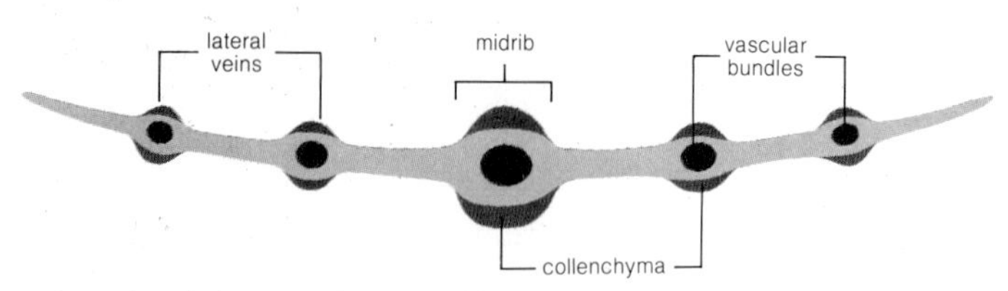

collenchyma (*n*) tissue (p. 88) of cells with thick cellulose (p. 17) cell walls (p. 17), especially at the angles of the cells, found in the stems of many herbs (p. 136), and in leaves. Its function is support.
aerenchyma (*n*) tissue (p. 88) with air-filled spaces between its cells, common in aquatic (p. 161) plants.
pith (*n*) tissue (p. 88), sometimes soft, in the centre of the stem of a non-woody (↓) dicotyledon (p. 131). Its function is to store food.
axis (*n*) general term for any stalk, stem, or long central organ (p. 88), from which other organs grow, e.g. the trunk (↓) of a tree.
trunk (*n*) the main woody stem of a tree, consisting of heartwood (↓), sapwood (↓) and bark (p. 94).
buttress (*n*) a large, flattened woody structure, growing outwards and downwards from near the base of the trunk (↑) of a tree. Buttresses are found especially in very large trees in tropical (p. 162) rain forests (p. 158).
bole (*n*) the trunk (↑) of a tree.
branch (*n*) a lateral (↓) shoot on a main axis (↑), e.g. the trunk (↑) of a tree.
lateral (*adj*) at, on, or of the side.
architecture (*n*) the way in which the branches of a tree are arranged on the trunk (↑), and the way in which the vegetative (p. 60) and reproductive (p. 59) axes (↑) are arranged on the branches.
crown (*n*) the top of a tree, including the branches and leaves.

parts of a tree

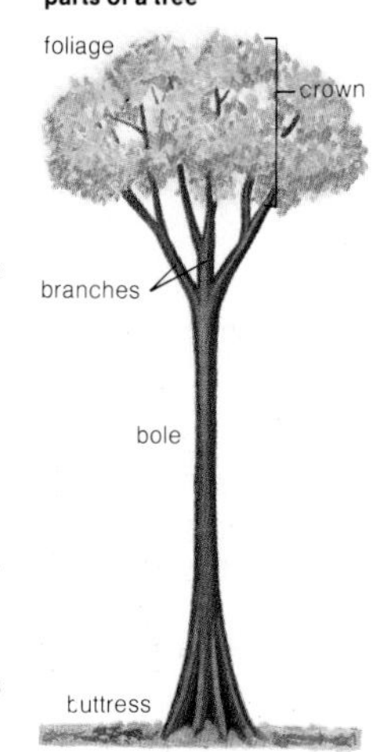

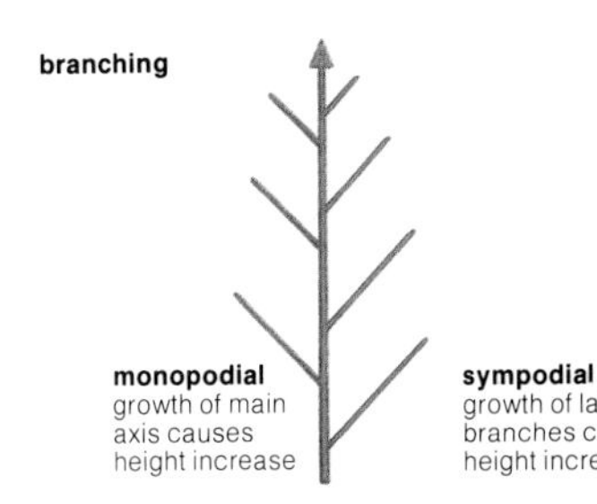

sympodial (*adj*) of a kind of growth in which the main axis (↑) of the plant is formed by the growth of lateral (↑) buds (p. 110) near the apex (p. 90) of the shoot, instead of by continuous growth from the apex.

monopodial (*adj*) of a kind of growth in which the main axis (↑) of the plant is formed by continuous growth of the same shoot apex (p. 90), with lateral (↑) branches arising from it.

dichotomous (*adj*) dividing equally into two, especially of branches.

orthotropic (*adj*) of axes (↑) growing upwards.

plagiotropic (*adj*) of branches which grow more or less parallel (p. 110) to the ground.

wood (*n*) hard tissue (p. 88) made of the remains of dead xylem (p. 106) cells in the stems of perennial (p. 117) plants. Wood contains lignin (↓), and its function is to support the plant and to conduct water. **woody** (*adj*).

lignin (*n*) a complex aromatic (p. 31) compound which is deposited in the cellulose (p. 17) cell walls (p. 17) of the xylem (p. 106) and sclerenchyma (p. 91) during the process of secondary thickening (p. 94). Wood is made mostly of lignin. **lignify** (*v*), **lignified** (*adj*).

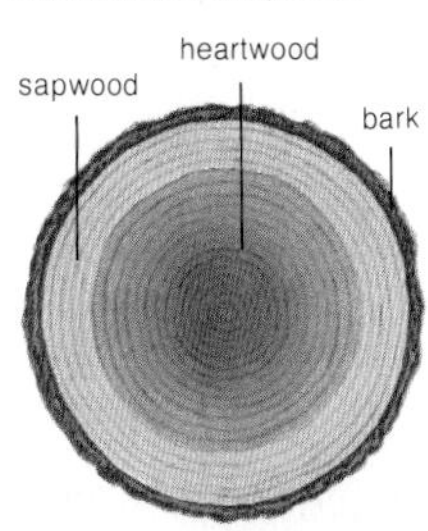

sapwood (*n*) the outer part of the xylem (p. 106) of a stem, containing some living cells. The sapwood lies outside the heartwood (↓), and its main function is translocation (p. 101).

heartwood (*n*) the wood in the centre of a trunk (↑) or branch. Heartwood is usually dense and compact, and it helps to support the tree. It is often darker than the outer wood, and is unable to conduct sap (p. 102).

primary thickening the thickening of a stem or root which occurs near the growing apex (p. 90).

secondary thickening the thickening of a stem or root because of the activity of the cambium (p. 108) to give xylem (p. 106) and phloem (p. 108). It provides the plant with extra support and vascular (p. 122) tissue.

pachycaul (*adj*) having fat stems due to massive primary thickening (↑), e.g. in many palms (p. 130). **pachycauly** (*n*).

leptocaul (*adj*) having thin stems, without heavy primary thickening (↑), as in most trees. **leptocauly** (*n*).

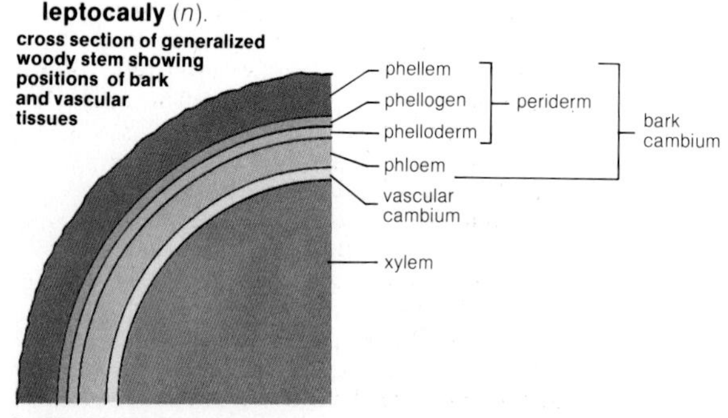

bark (*n*) tissue (p. 88) usually made of dead cork (↓) cells and phloem (p. 108), on the outside of woody stems. Its function is to protect the stem.

cork (*n*) tissue (p. 88) of dead cells with suberin (↓) cell walls (p. 17), forming part of the bark (p. 94).

periderm (*n*) tissue (p. 88) forming part of the bark (p. 94), consisting of phelloderm (↓), phellogen (↓) and phellem (↓).

phellem (*n*) = cork (↑).

phelloderm (*n*) the inner layer of the periderm (↑), inside the cork (↑).

phellogen (*n*) the cambium (p. 108) which produces cork (↑) and phelloderm (↑). It is sometimes called the cork cambium.

suberin (*n*) a mixture of substances formed from fatty acids (p. 31), which is found in cork (↑) cell walls (p. 17). Suberin prevents water from passing through the cork.

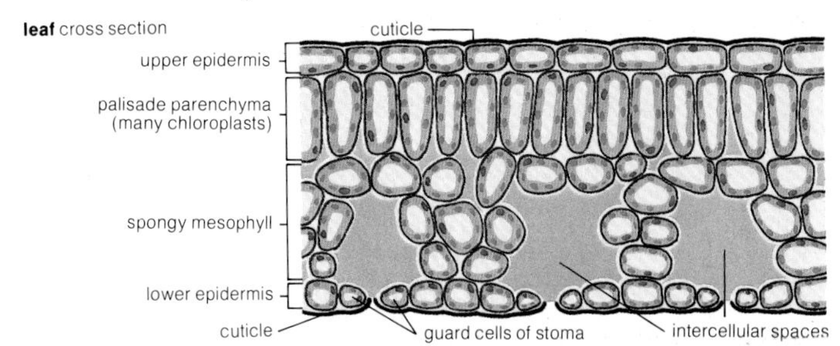

leaf (*n*) the plant organ (p. 88) whose function is photosynthesis (p. 32) and transpiration (p. 101). Leaves are produced from buds (p. 110) on the stem. Leaves have a wide range of form, but they nearly all share the inability to produce new growth from their apices (p. 90). In perennial (p. 117) plants, old leaves are replaced by new ones as the plant grows.

mesophyll (*n*) the tissue (p. 88) between the upper and lower epidermis (p. 90) of a leaf. In dicotyledons (p. 131) it is differentiated (p. 110) into palisade parenchyma (↓) and spongy mesophyll (↓), but in most monocotyledons (p. 130) it is undifferentiated.

spongy mesophyll a tissue (p. 88) in the leaves of many plants, e.g. dicotyledons (p. 131), lying underneath the palisade parenchyma (↓). It is composed of large cells with many intercellular spaces (↓) between them.

intercellular space the spaces between cells. In some tissues (p. 88), e.g. the spongy mesophyll (↑) of leaves, the intercellular spaces are large and filled with air.

palisade parenchyma layer of upright cells below the upper epidermis (p. 90) of leaves, especially in dicotyledons (p. 131). The cells are rich in chloroplasts (p. 32), and their main function is photosynthesis (p. 32).

cuticle (*n*) layer of cutin (p. 96) on the surface of leaves and green stems, which prevents evaporation (p. 12) and protects the plant against attack from herbivores (p. 153) and pathogens (p. 144).

cutin (*n*) substance made of fatty acid (p. 31) products, which is impermeable (p. 102) to water.
wax (*n*) substance covering the surfaces of many plants. It is composed of a variety of organic (p. 11) compounds and polymers (p. 10), many of which are derived from lipids (p. 31). Wax coverings help to reduce the evaporation (p. 12) of water from leaves, and can also reflect light.
stoma (*n*) a pore (p. 19) in the surface of a leaf, usually consisting of two guard cells (↓) with a space between them. Stomata (*pl.*) can be opened and closed, controlling the evaporation (p. 12) of water from the leaf, and the entry of CO_2 into the leaf.
guard cells the pair of cells forming a stoma (↑).
foliage (*n*) the leaves of a plant collectively.
hypodermis (*n*) an extra layer of protective cells beneath the epidermis (p. 90) in the leaves, stems and roots of some plants.

stomata
surface view of leaf

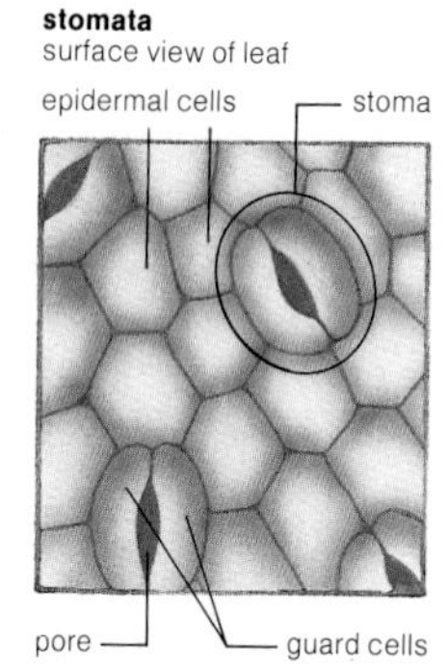

leaf

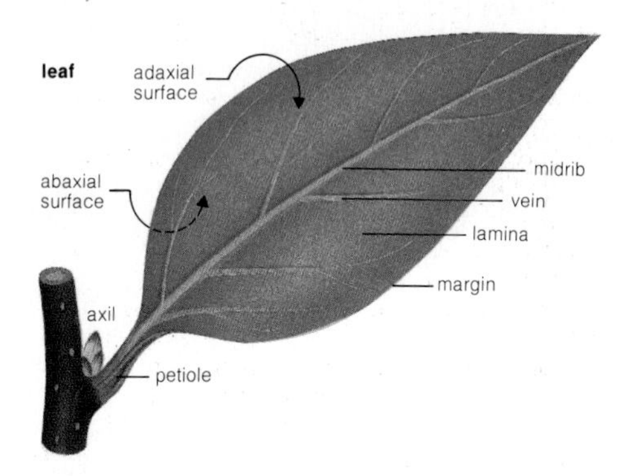

petiole (*n*) the stalk of a leaf, which joins it to a node (p. 90) on the stem.
axil (*n*) the point where the upper side of the petiole (↑) of a leaf joins the stem. **axillary** (*adj*).
lamina (*n*) the parts of the leaf on either side of the midrib (↓).
blade (*n*) either the whole leaf, excluding the petiole (↑), or all the parts of the leaf except the midrib (↓), i.e. the lamina (↑).

midrib (*n*) the central vein (↓) of a leaf.
vein (*n*) one of the many lines which can be seen on the surface of a leaf, marking the position of the vascular bundle (p. 105).

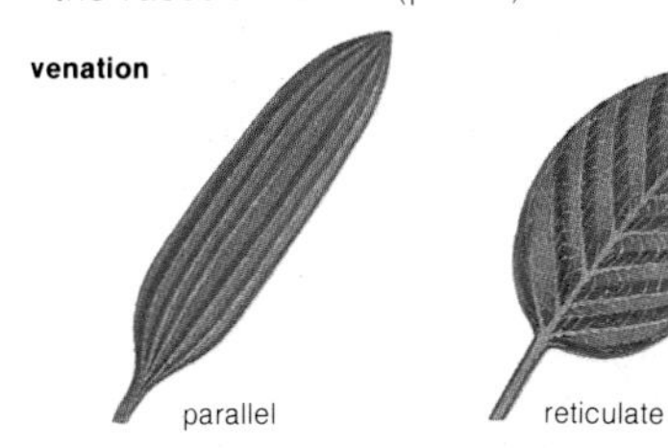

venation (*n*) the pattern of veins (↑) on the surface of a leaf. In most dicotyledons (p. 131) the venation is reticulate (↓), and in most monocotyledons (p. 130) it is parallel (p. 110).
reticulate (*adj*) of the veins (↑) of leaves, when their pattern is like a network.
margin (*n*) the edge, e.g. of a leaf.
adaxial (*adj*) on the top side of a leaf, that is, pointing towards the stem.
abaxial (*adj*) on the underside of a leaf, that is, pointing away from the stem.
simple (*adj*) of leaves which are not divided into leaflets (p. 98).
entire (*adj*) of leaves without lobes (↓).
digitate (*adj*) of leaves in which the lamina (↑) is divided like the fingers of a hand.
dissected (*adj*) of leaves which have many lobes (↓).
lobe (*n*) a flat, roundish piece of tissue (p. 88), as at the margin (↑) of a digitate (↑) or dissected (↑) leaf. Also, petals (p. 70) are sometimes called corolla (p. 70) lobes.

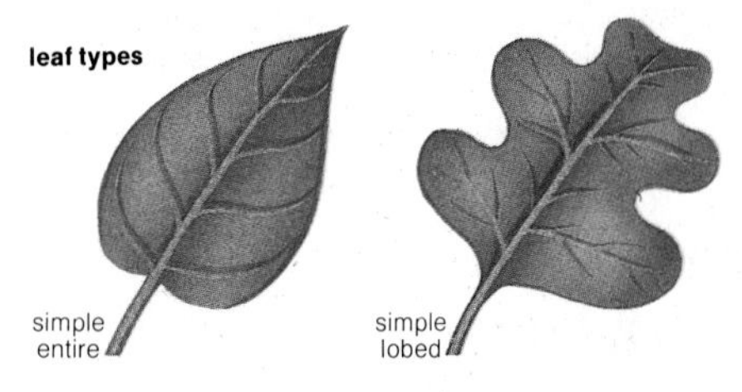

compound[2] (*adj*) of leaves which are divided into several or many leaflets (↓) without axillary (p. 96) buds (p. 110).

leaflet (*n*) one of the small leaf-like divisions of a compound (↑) leaf.

rachis (*n*) the main axis (p. 92) of a pinnately (↓) compound (↑) leaf. The rachis is a continuation of the petiole (p. 96).

rhachis (*n*) = rachis (↑).

palmate (*adj*) of compound (↑) leaves with leaflets (↑) arising from a central point at the end of the petiole (p. 96), or, of simple (p. 97) leaves with lobes (p. 97), in which the main veins (p. 97) arise in the same way.

pinnate (*adj*) of a compound (↑) leaf with a central axis (p. 92) and leaflets (↑) - pinnae (↓) - on either side of it.

pinna (*n*) the leaflet (↑) of a pinnately (↑) compound (↑) leaf. **pinnae** (*pl.*).

pinnule (*n*) the leaflet (↑) on the pinna (↑) of a bipinnate (↓) leaf, as in many members of the order (p. 134) Filicales, the ferns (p. 126).

bipinnate (*adj*) of pinnate (↑) leaves with their pinnae (↑) divided into pinnules (↑), as in many ferns (p. 126).

phyllotaxy (*n*) the arrangement of leaves on a stem, e.g. opposite (↓) phyllotaxy, alternate (↓) phyllotaxy, spiral (↓) phyllotaxy, whorled (↓) phyllotaxy. This is an important character in classification (p. 132).

spiral (*adj*) of nodes (p. 90) and leaves which are arranged on the stem in a helical (p. 51) way, or, of the thickening of the walls of xylem (p. 106) cells.

whorl (*n*) a group of three or more organs (p. 88) of the same kind, arising at the same level on a stem and arranged in a circle, e.g. the petals (p. 70) of a flower or the branches of a horsetail (p. 127). **whorled** (*adj*).

alternate (*adj*) of an arrangement of leaves which arise singly on a stem, each one on the other side of the stem from the leaf below or above it.

opposite (*adj*) of an arrangement of two leaves which arise at the same node (p. 90), on either side of the stem.

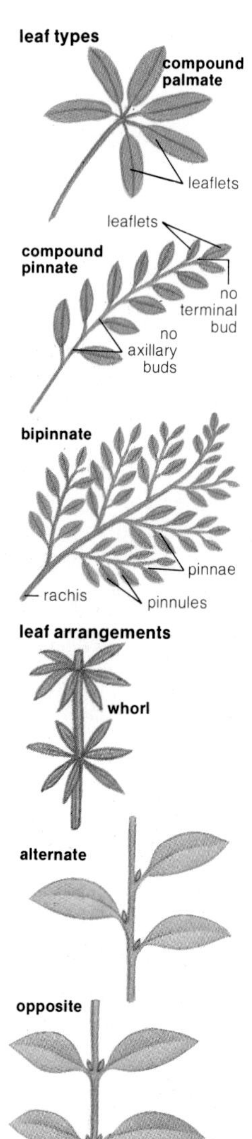

rosette

variegated leaves

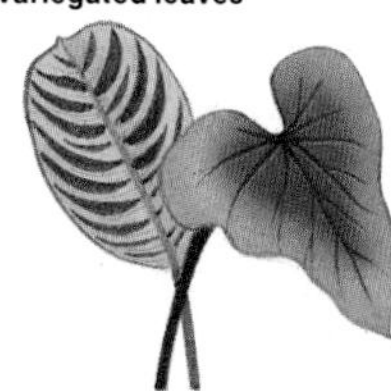

drip tip

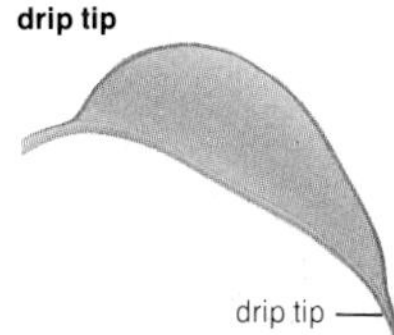

needles

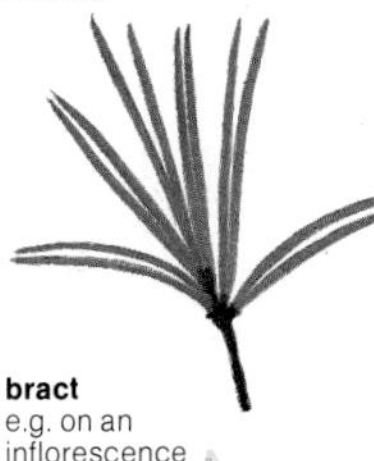

bract e.g. on an inflorescence

bracts

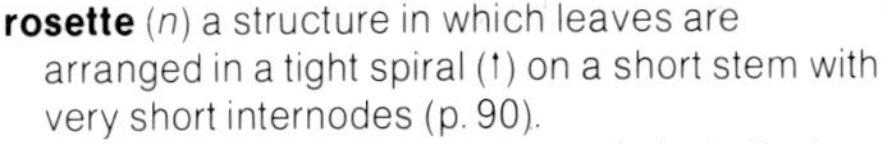

rosette (*n*) a structure in which leaves are arranged in a tight spiral (↑) on a short stem with very short internodes (p. 90).

succulent (*adj*) of plants or parts of plants that are thick and fleshy (↓), owing to the presence of water-storing tissues (p. 88), e.g. in the cactus family (p. 134), Cactaceae.

fleshy (*adj*) of organs (p. 88) which are thick and often juicy.

coriaceous (*adj*) of leaves which are thick and stiff, like leather.

chartaceous (*adj*) of leaves which are like thick paper.

membranaceous (*adj*) of leaves which are very thin.

variegated (*adj*) of leaves with patches of different colour. **variegation** (*n*).

heterophyllous (*adj*) of plants which have two different kinds of leaves, e.g. when the leaves of the young plant are different from the leaves of the old plant, as in many species (p. 134) of the ivy family (p. 134), Araliaceae. **heterophylly** (*n*).

phyllode (*n*) a flat petiole (p. 96), which has the appearance of a leaf.

drip tip a long pointed tip to the leaf, which helps water to run off the leaf surface. Drip tips are common in wet tropical (p. 162) forests (p. 158).

needle (*n*) the long thin leaf of some conifers (p. 128).

bract (*n*) a small leaf, with a flower or part of an inflorescence (p. 80) growing from its axil (p96).

bracteole (*n*) a small bract (↑).

stipule (*n*) a small, leaf-like organ (p. 88), found in many plants, which grows at the base of a petiole (p. 96), sometimes protecting an axillary (p. 96) bud (p. 110).

exstipulate (*adj*) without stipules (↑).

stipule

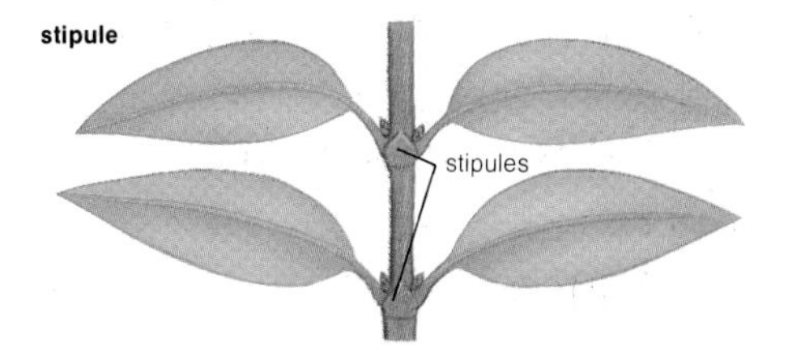

sheath (*n*) a protective covering, e.g. the lower part of the leaf of a grass, which is rolled around the stem.
coleoptile (*n*) a sheath (↑) which protects the young shoot tip in grasses (p. 130).
auricle (*n*) a small outgrowth at the side of the base of the leaf in some grasses (p. 130).
ligule[2] (*n*) a thin flap of tissue (p. 88) at the top of the leaf-sheath (↑) in many grasses (p. 130).
spine (*n*) a long, thin, sharp, stiff organ (p. 88) on the surface of the stems, and also sometimes of the leaves, of some plants, which is a defence against attack by herbivores (p. 153).
thorn (*n*) a sharp, pointed outgrowth on the surface of a plant, especially on the stem. Thorns can be simple outgrowths of the epidermis (p. 90), or can be modifications of other organs (p. 88), for example, the stipules (p. 99).
armed (*adj*) having thorns (↑) or spines (↑).
scale (*n*) a small outgrowth, e.g. on the petiole (p. 96) of a fern (p. 126) frond (p. 126).
trichome (*n*) a hair on the epidermis (p. 90) of a plant.
pubescent (*adj*) = hairy.
tomentose (*adj*) having a thick covering of very short hairs.
indumentum (*n*) the hairy part of a plant.
sessile (*adj*) of organs (p. 88) without a stalk, e.g. a leaf which has no petiole (p. 96) and is attached directly to the stem.

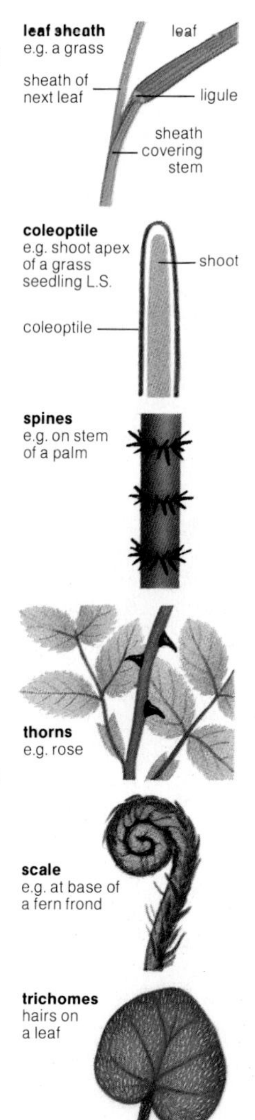

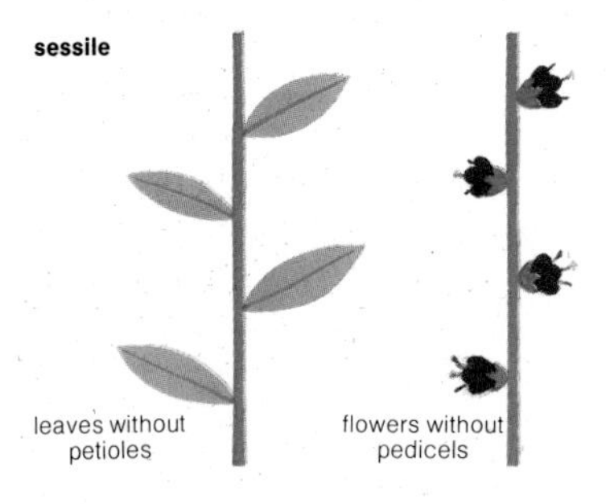

translocation (*n*) the movement of substances in the vascular system (p. 105) from one part of a plant to another. **translocate** (*v*).

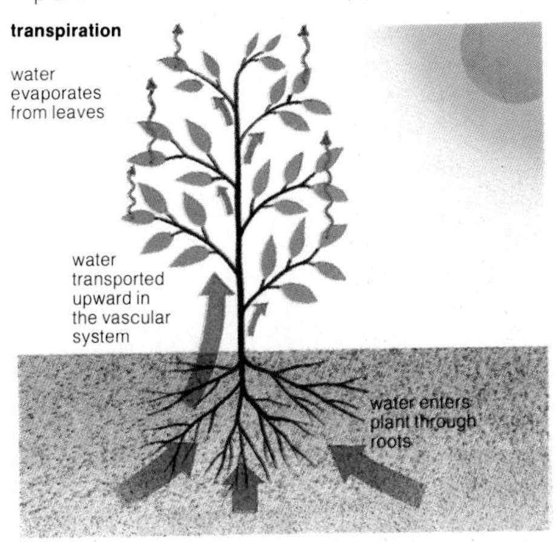

transpiration (*n*) the process of upward movement of sap (p. 102) in the xylem (p. 106), due to the water potential (p. 103) gradient (p. 24) caused by the evaporation (p. 12) of water from the leaves. **transpire** (*v*).

potometer (*n*) an instrument for measuring the rate of transpiration (↑).

transpiration stream the upward-moving sap (p. 102) in the xylem (p. 106).

evapotranspiration (*n*) the process of losing water from vegetation (p. 150), caused by the evaporation (p. 12) of water at the surfaces of leaves.

uptake (*n*) the process of taking water and nutrients (p. 111) from the soil into the roots of a plant, or of taking substances into a cell or organelle (p. 16).

active transport the movement of substances across membranes (p. 18), using energy. This is necessary when a substance is being transported from the side of the membrane where it is less concentrated (p. 12) to the side where it is more concentrated.

diffusion (*n*) the natural movement of molecules (p. 9) of a solute (p. 12) from regions of higher concentration (p. 12) to regions of lower concentration. **diffuse** (*v*).

apoplast (*n*) the non-living parts of a plant, i.e. the xylem (p. 106), the cellulose (p. 17) cell walls (p. 17) and the intercellular spaces (p. 95).

symplast and apoplast pathways

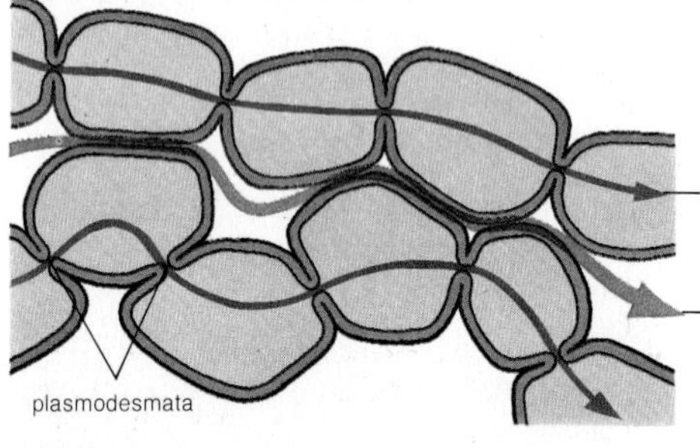

symplast (*n*) the living parts of a plant, i.e. the cells containing cytoplasm (p. 18).

root pressure the pressure which partly causes upward movement of xylem (p. 106) sap (↓), resulting from the active transport (p. 101) of solutes (p. 12) into the xylem which itself causes osmotic (↓) flow of water into the xylem.

sap (*n*) the water and nutrients (p. 111) contained and transported in the xylem (p. 106) or phloem (p. 108). Sap is also a general name for any liquid exuded (p. 112) from the plant when it is cut.

latex (*n*) a coloured, sticky or milky liquid, produced by specialized cells, which some plants exude (p. 112) when cut, e.g. rubber.

osmosis (*n*) the process by which water moves across semi-permeable (↓) membranes (p. 18) from a hypotonic (↓) solution (p. 12) to a hypertonic (↓) solution. **osmotic** (*adj*).

osmotic pressure the pressure needed to prevent osmotic (↑) movement of pure water across a semi-permeable (↓) membrane (p. 18) into a solution (p. 12). *See also* **osmotic potential** (↓).

osmosis

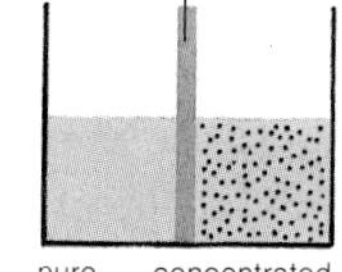

water diffuses across membrane until pressure from solution prevents further movement

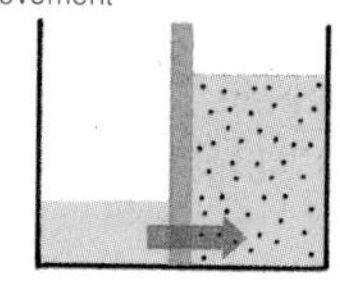

permeable (*adj*) of membranes (p. 18) which allow the movement of substances from one side to the other. **permeability** (*n*).

impermeable (*adj*) of membranes (p. 18) which do not allow the movement of substances from one side to the other.

semipermeable (*adj*) of membranes (p. 18) which let some substances pass through them, but not others. The membranes in plant cells are mostly permeable (↑) to small molecules (p. 9), e.g. water (H_2O), monosaccharides (p. 28) and amino acids (p. 56), but not to large molecules, e.g. polypeptides (p. 56).

hypotonic (*adj*) less concentrated (p. 12).

hypertonic (*adj*) more concentrated (p. 12).

isotonic (*adj*) of two solutions (p. 12) with the same concentration (p. 12) of solute (p. 12) and the same osmotic pressure (↑).

water potential a measure, expressed in units of pressure, of the chemical difference between pure water and a solution (p. 12) in which water is the solvent (p. 12). Water diffuses (↑) from a solution of high water potential to a solution of low water potential if they are separated by a semi-permeable (↑) membrane (p. 18). In turgid (p. 104) plant cells, the water potential is equal to the sum of the osmotic pressure (↑), and matric potentials (↓).

osmotic potential a measure of the pressure that needs to be applied to a solution (p. 12) in order to make its water potential (↑) equal to that of pure water. When this pressure is applied to a solution, pure water will not pass into it across a semi-permeable (↑) membrane (p. 18). When there is no matric potential (↓) or pressure potential (↓), i.e. under experimental conditions, osmotic potential is equal to water potential (↑).

pressure potential a measure, in units of pressure, of the squeezing effect that the cell wall (p. 17) has on the contents of the cell, in turgid (p. 104) plant cells.

matric potential a measure, in units of pressure, of the attraction between water molecules (p. 9) and the organic (p. 11) compounds of the cell wall (p. 17) and cell.

plasmolyzed cell

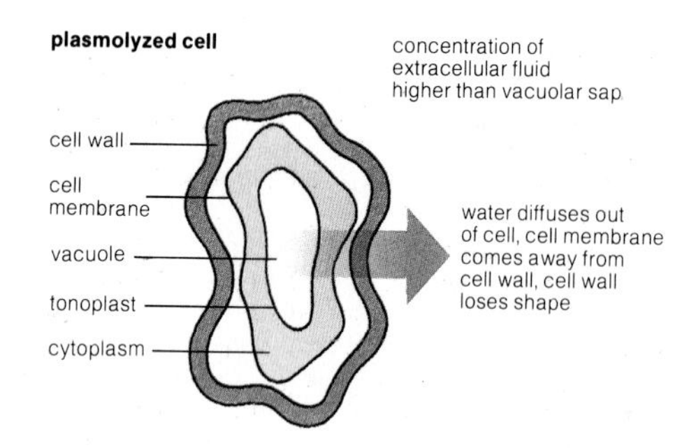

plasmolysis (*n*) the separation of the plasmalemma (p. 18) from the cell wall (p. 17) and the shrinking of the protoplast (p. 18) which occurs when water passes out of the cell due to the presence of a hypertonic (p. 103) solution (p. 12) outside it.

turgor (*n*) the tension on a cell wall (p. 17) due to the pressure of water inside the cell.

turgid cell

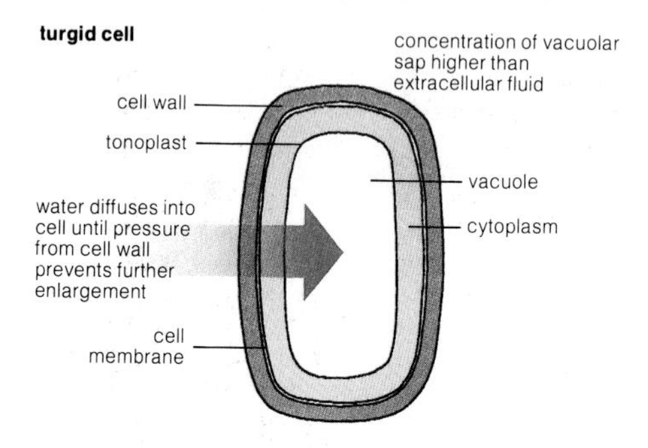

turgid (*adj*) of the state of a cell which can absorb no more water by osmosis (p. 102), because the cell wall (p. 17) prevents an increase in size.

wilt (*v*) to droop (of leaves and green stems of a plant) either because the rate of evaporation (p. 12) of water from the leaves is greater than the rate of uptake of water by the roots, or because of disease.

vascular cylinder

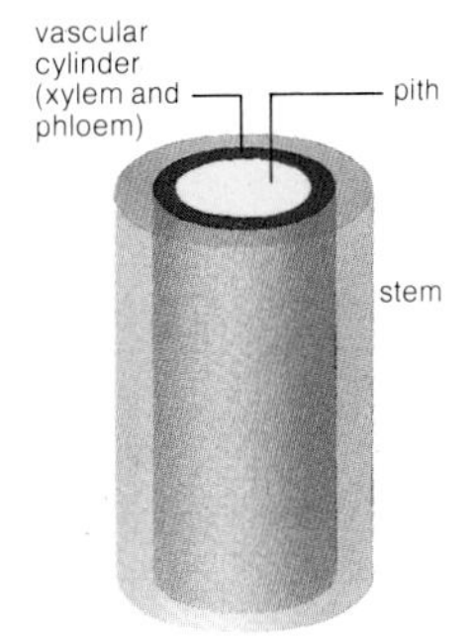

vascular system the tissues (p. 88) consisting of xylem (p. 106) and phloem (p. 108) cells, which translocate (p. 101) substances from one part of a plant to another. The development of vascular systems has made it possible for plants to evolve (p. 139) on land.

vascular cylinder a tube of vascular (p. 122) tissue (p. 88) consisting of xylem (p. 106) and phloem (p. 108) in a root or stem.

stele (*n*) the vascular cylinder (↑) of a stem or root, including the pith (p. 92) if this is present.

vascular bundle a thread of vascular (p. 122) tissue (p. 88) in the vein (p. 97) of a leaf, or in a stem.

vascular bundle

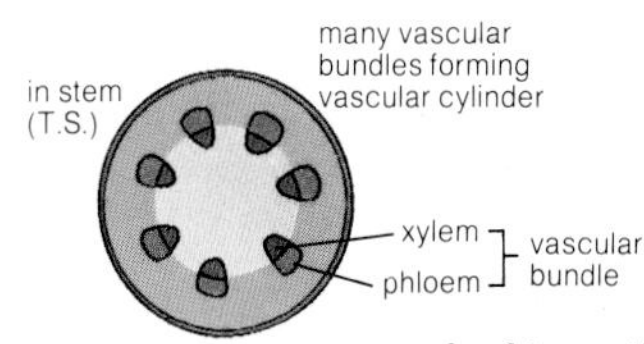

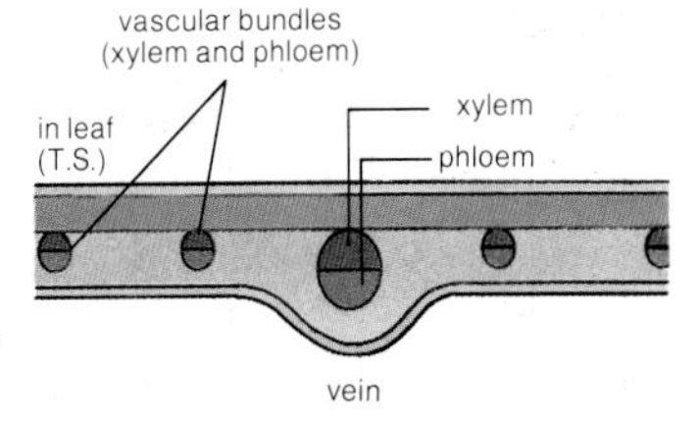

leaf trace the vascular (p. 122) tissue (p. 88) which branches off from the stem, at a node (p. 90), into a leaf.

leaf traces and gaps at the node of a stem

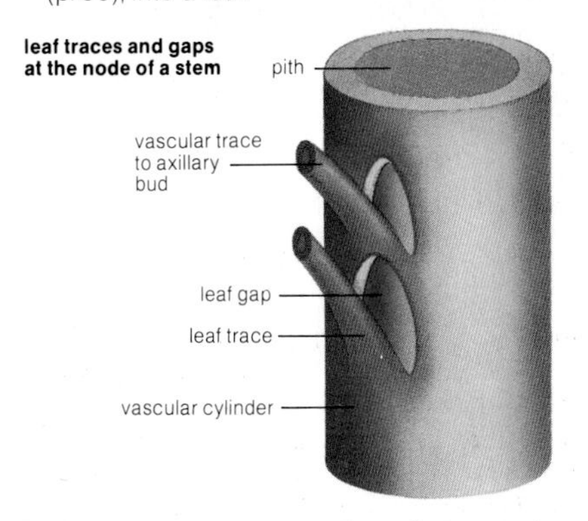

leaf gap a gap in the vascular cylinder (↑) of a stem, just above a node (p. 90).

megaphyll (*n*) a leaf whose leaf trace (p. 105) makes a gap in the vascular system (p. 105) of the stem.

microphyll (*n*) a leaf whose leaf trace (p. 105) does not make a gap in the vascular system (p. 105) of the stem.

bundle sheath a layer of cells around the vascular bundle (p. 105) in a leaf.

xylem (*n*) tissue (p. 88) in the vascular system (p. 105) of a plant, consisting of tracheids (↓), vessels (↓), parenchyma (p. 90) and sclerenchyma (p. 91). The vessels, tracheids and sclerenchyma have lignified (p. 93) cell walls (p. 17). Most xylem cells are dead, and contain no cytoplasm (p. 18). The function of the xylem is to translocate (p. 101) water and nutrients (p. 111) from the roots to the stems and leaves.

bundle sheath
T.S. leaf of a C4 plant

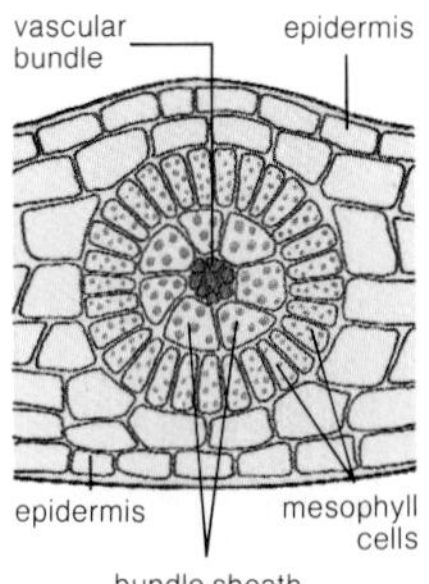

position of xylem and phloem in young and old roots

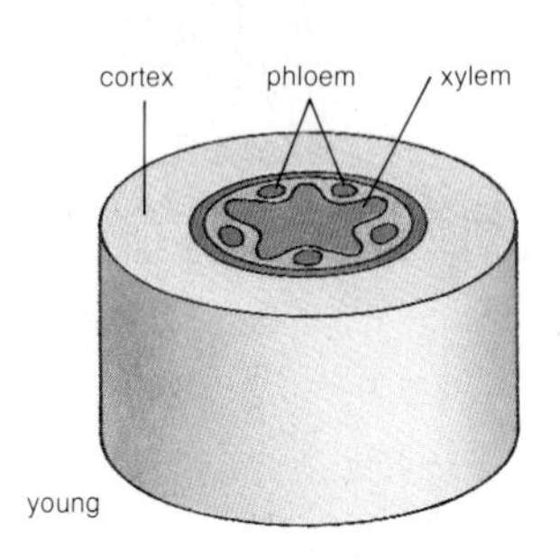

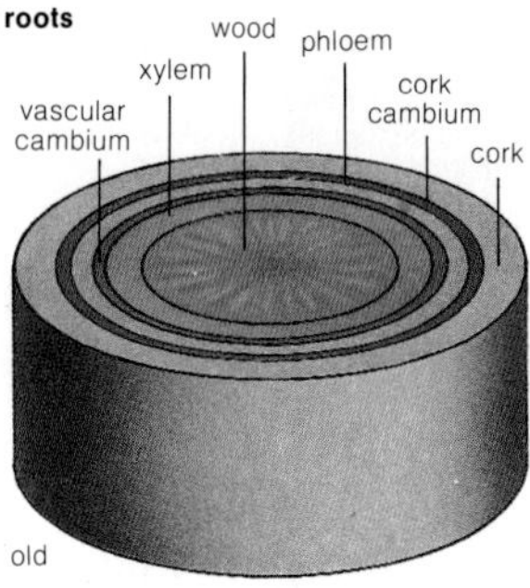

position of xylem and phloem in young and old stems

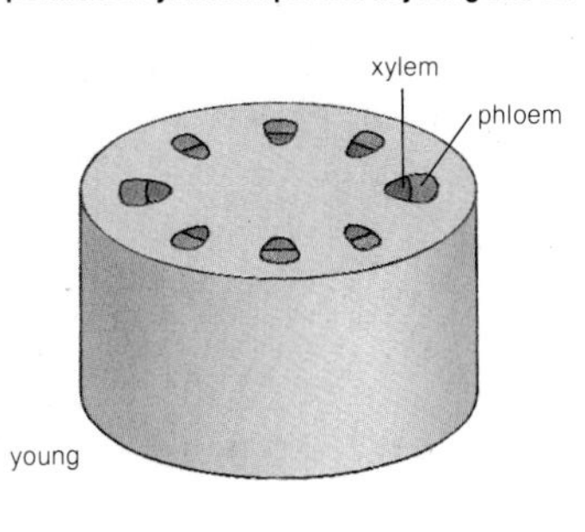

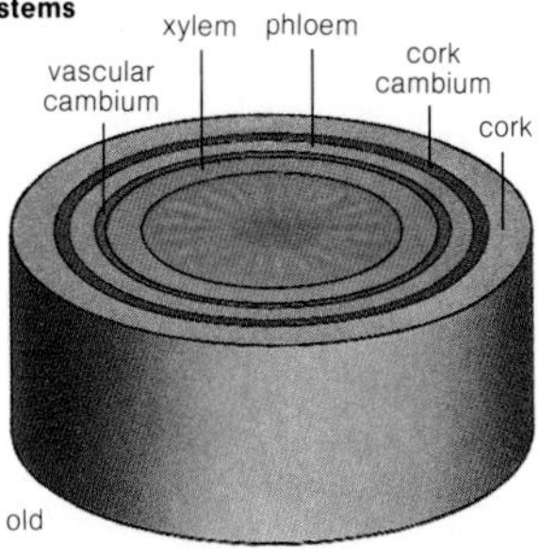

cell types in xylem

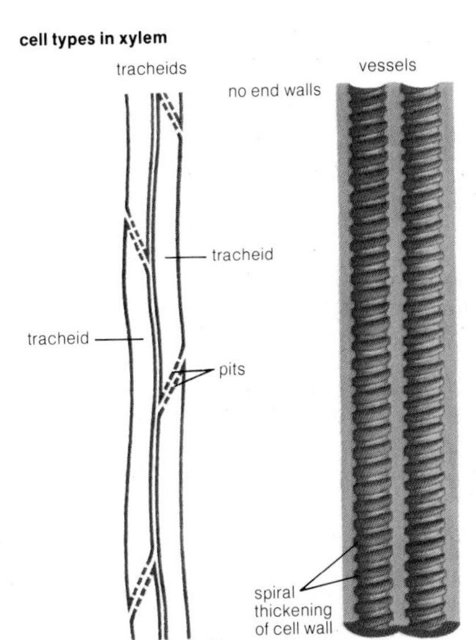

vessel (*n*) conducting tissue (p. 88) in the xylem (↑), consisting of vessel elements (↓), found mainly in angiosperms (p. 130).

vessel element a long, often thin, dead cell in the vessels (↑) of the xylem (↑). Vessel elements are arranged end to end, with large holes in the end walls, through which the xylem sap (p. 102) can pass. The cell walls (p. 17) of vessel elements are thickened with lignin (p. 93).

tracheid (*n*) a long, thin, dead cell in the xylem (↑), with closed ends and lignified (p. 93) walls. Xylem sap (p. 102) passes from one tracheid to the next through pits (↓) in the cell walls (p. 17).

scalariform (*adj*) of tracheids (↑) and vessels (↑) whose call walls (p. 17) have ladder-like ridges of thickening.

pit (*n*) an unthickened point in the cell wall (p. 17), which is usually next to a pit in the cell wall of the next cell. Pits enable the easy passage of substances from one cell to the next. They are common in the tracheids (↑) of the xylem (↑).

phloem (*n*) one of the conducting tissues (p. 88) in the vascular system (p. 105). Phloem, unlike xylem (p. 106), is mainly a living tissue, whose cells contain cytoplasm (p. 18). It is made of sieve elements (↓) and companion cells (↓). The phloem can translocate (p. 102) substances in both directions, and its main function is to translocate the products of photosynthesis (p. 32) from leaves to other parts of the plant.

sieve-tube (*n*) the tissue (p. 88) in the phloem (↑) through which substances are translocated (p. 102). It consists of sieve elements (↓) with sieve plates (↓) between them.

sieve element a cell in the sieve-tube (↑) of the phloem (↑). Sieve elements are long, thin, living cells with thin cell walls (p. 17) and sieve plates (↓) at their ends. Translocation (p. 102) of substances takes place in the sieve elements.

sieve plate the wall at the end of a sieve element (↑), which has large pores (p. 19) through which substances can pass. Sieve plates contain callose (↓).

callose (*n*) a carbohydrate (p. 28) polymer (p. 10) which is found in sieve plates (↑), pollen tubes (p. 74) and on injured surfaces.

callus[2] (*n*) a tissue (p. 88) produced on injured plant surfaces. Callus tissue contains callose (↑).

companion cell a small, living cell next to the sieve elements (↑) in the phloem (↑).

vascular cambium

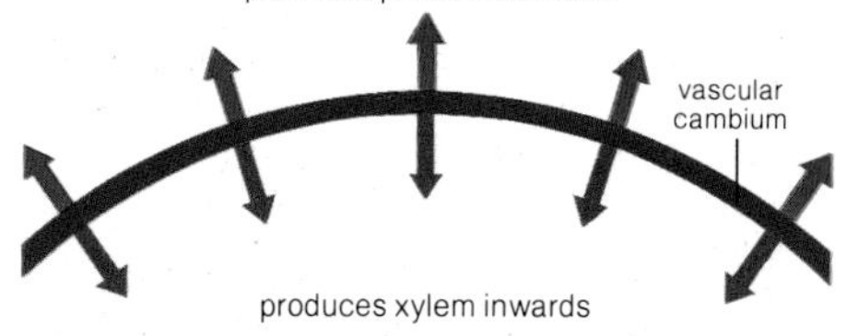

cambium (*n*) a meristem (↓) in the vascular system (p. 105). In perennial (p. 117) plants, it produces a new layer of vascular (p. 122) tissue (p. 88) each year, producing xylem (p. 106) on the inside and phloem (↑) on the outside.

phloem
(L.S.)

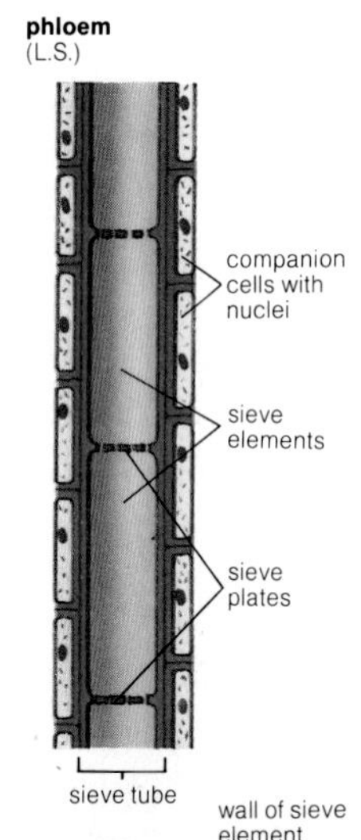

growth (*n*) all the processes in an organism which result in an increase in size. The ability to grow is one of the important characteristics of living organisms.

vegetative growth growth of the tissues (p. 88) and organs (p. 88) not involved in sexual (p. 59) reproduction (p. 59). Vegetative growth occurs by mitosis (p. 45) and the lengthening and enlargement of cells.

development (*n*) the changes of structure and appearance of new organs (p. 88) and tissues (p. 88) as an organism grows. **develop** (*v*).

ontogeny (*n*) the process of development of an individual (p. 135) from zygote (p. 61) to adult.

morphogenesis (*n*) the development of shape and structure of organs (p. 88) and tissues (p. 88).

meristems

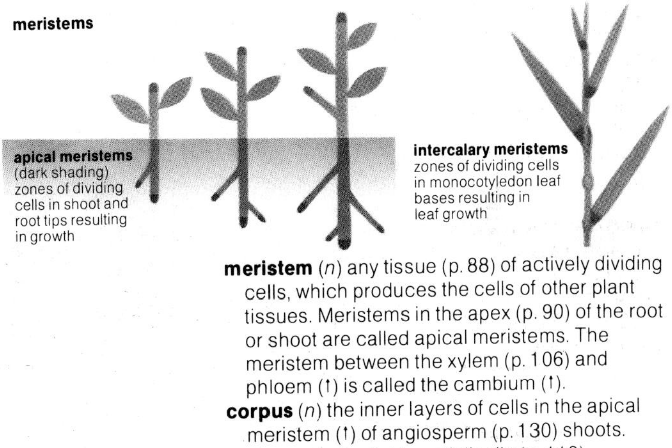

apical meristems (dark shading) zones of dividing cells in shoot and root tips resulting in growth

intercalary meristems zones of dividing cells in monocotyledon leaf bases resulting in leaf growth

meristem (*n*) any tissue (p. 88) of actively dividing cells, which produces the cells of other plant tissues. Meristems in the apex (p. 90) of the root or shoot are called apical meristems. The meristem between the xylem (p. 106) and phloem (↑) is called the cambium (↑).

corpus (*n*) the inner layers of cells in the apical meristem (↑) of angiosperm (p. 130) shoots. Corpus cells divide anticlinally (p. 110), producing the inner tissues (p. 88) of the shoot.

tunica (*n*) the outer layer or layers of cells in the apical meristem (↑) of angiosperm (p. 130) shoots. Tunica cells divide periclinally (p. 110), producing the surface tissues (p. 88) of the shoot.

intercalary (*adj*) of the meristems (↑) at the base of leaves and stems in monocotyledons (p. 130).

periclinal (*adj*) of cell divisions (p. 45) which occur parallel (↓) to the surface of the plant.
anticlinal (*adj*) of cell divisions (p. 45) which occur at right angles to the surface of the plant.
parallel (*adj*) of lines or planes running in the same direction and never meeting.
primordium (*n*) an undeveloped organ (p. 88), e.g. a leaf bud (↓) contains leaf primordia (*pl.*), a young flower bud contains the primordia of reproductive (p. 59) organs.
plastochrone (*n*) the time between the formation of one leaf primordium (↑) and the next.
bud (*n*) an undeveloped shoot covered with protecting scales (p. 100), consisting of a very short axis (p. 92) bearing primordia (↑) of leaves or flower parts.

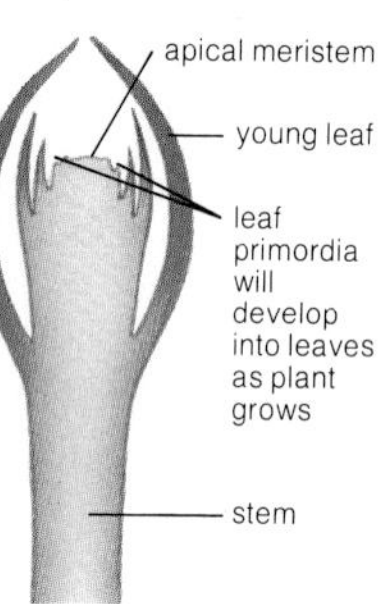

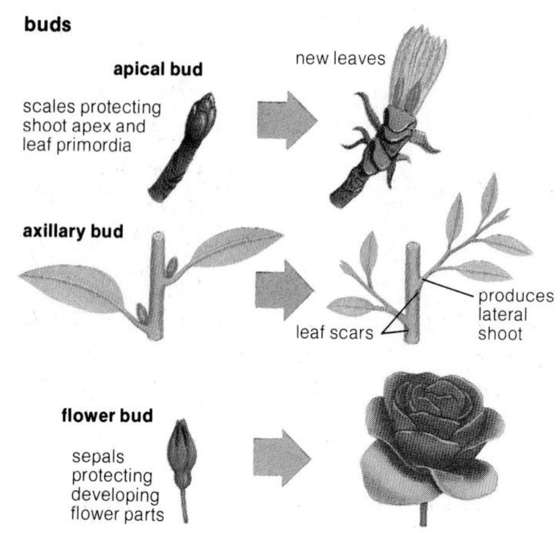

differentiated (*adj*) of cells which have developed a particular structure in relation to their function in a tissue (p. 88) or organ (p. 88). **differentiate** (*v*), **differentiation** (*n*).
undifferentiated (*adj*) of cells in an embryo (p. 85) or young part of a plant, e.g. a meristem (p. 109), which are all the same and have not developed into different tissues (p. 88). In many simple plants, e.g. the prothalli (p. 122) of ferns (p. 126), most of the cells are undifferentiated.

physiology (*n*) the study of the internal processes of organisms.

regeneration[1] (*n*) (1) the growth of new tissue (p. 88) on a part of a plant that has been damaged; (2) the growth of new plants from perennating (p. 117) organs (p. 88), e.g. rhizomes (p. 60). **regenerate** (*v*).

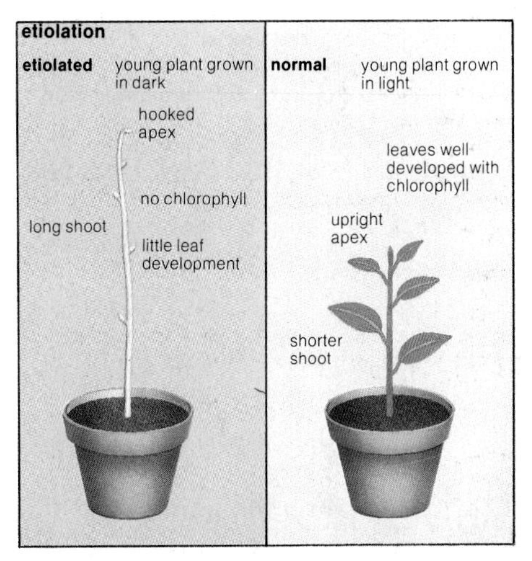

etiolation (*n*) the process of rapid growth, without the production of chlorophyll (p. 36), that occurs in shoots kept in the dark. Etiolated shoots are long, thin and pale, and their leaves are very small. **etiolate** (*v*).

nutrition (*n*) the process of taking up nutrients (↓) and using them in metabolism (p. 14).

nutrient (*n*) an inorganic (p. 11) substance which plants require for growth. Nutrients are taken up from the soil by the roots, e.g. nitrate (p. 13), phosphate (p. 13).

trace element an element (p. 8) required by a plant in very small amounts, e.g. boron, molybdenum.

deficiency (*n*) the lack of a nutrient (↑) required for growth and development. Deficiency can lead to poor growth and disease.

secretion (*n*) the transport of a dissolved (p. 12) substance produced by a cell or organ (p. 88) out of that cell or organ. **secrete** (*v*).
excretion (*n*) the process of removing waste products of metabolism (p. 14) from a cell or organism. **excrete** (*v*).
gland (*n*) a group of cells on the surface of a plant, whose function is to secrete (↑) or excrete (↑) substances. **glandular** (*adj*).
exude (*v*) to secrete (↑) liquid from pores (p. 19), e.g. in guttation (↓), or from a cut surface.
exudate (*n*) the liquid exuded (↑) from pores (p. 19) and glands (↑) such as hydathodes (↓).
hydathode (*n*) a gland (↑) on the leaves of some plants, which exudes (↑) water.
guttation (*n*) the process of exuding (↑) sap (p. 102) or water through hydathodes (↑).
hormone (*n*) a substance which, in very small amounts, controls growth and development. Hormones are chemical messengers which are usually produced in one organ (p. 88) and transported to another part of the plant, where they have their effects. The five main groups of plant hormones are auxins (↓), gibberellins (↓), cytokinins (↓), ethene (p. 114) and abscisic acid (↓).

guttation

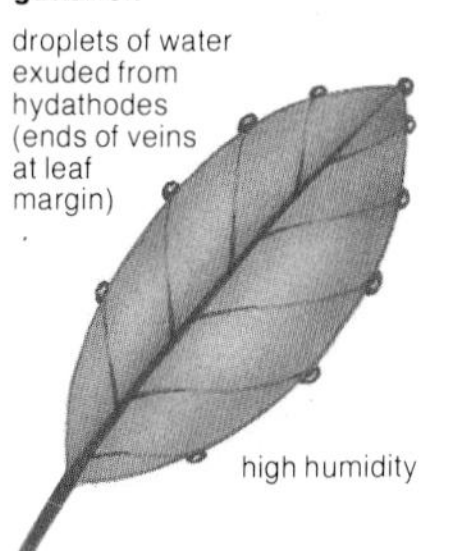

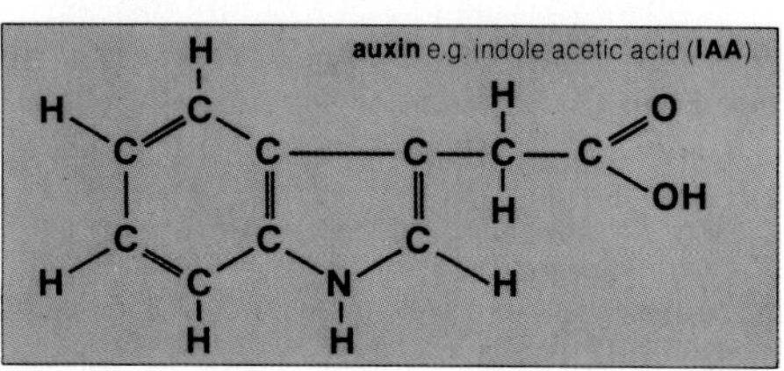

auxin (*n*) general name for an important group of plant hormones (↑). Indole acetic acid, IAA (↓) is the most common auxin. Auxins affect many processes, e.g. tropisms (p. 115), fruit growth, apical dominance (p. 114), stem growth. They can also inhibit (p. 14) root growth.
IAA = indole acetic acid (↓).
indole acetic acid the most common auxin (↑). Indole acetic acid is produced in the shoot apex (p. 90).

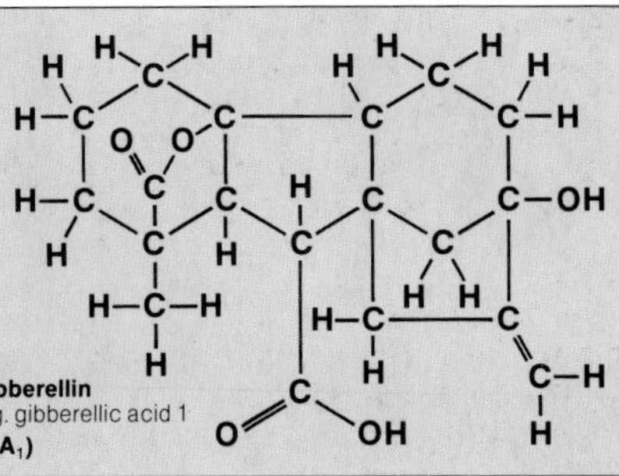

gibberellin
e.g. gibberellic acid 1
(GA_1)

gibberellins (*n.pl.*) group of chemically complex plant hormones (↑), important in the control of tropisms (p. 115), the lengthening of cells during growth, germination (p. 87) and other processes.

cytokinins
e.g. kinetin

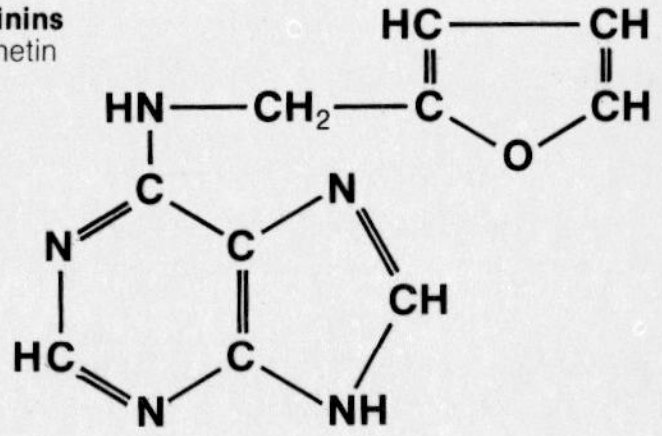

cytokinins (*n.pl.*) group of plant hormones (↑) which control cell division (p. 45).

abscisic acid a plant hormone (↑) which inhibits (p. 14) root growth and germination (p. 87), and which is important in the control of leaf abscission (p. 114).

abscisic acid

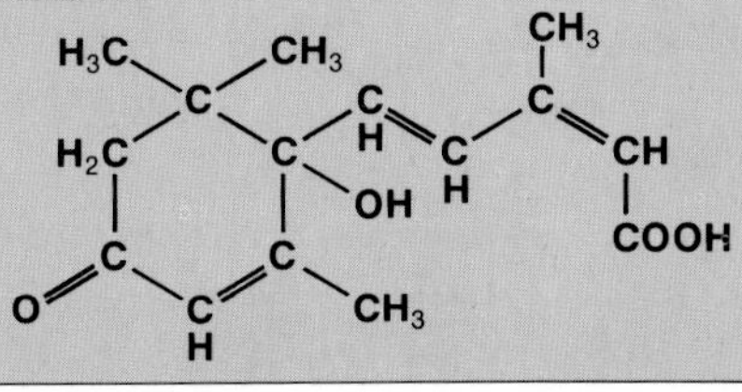

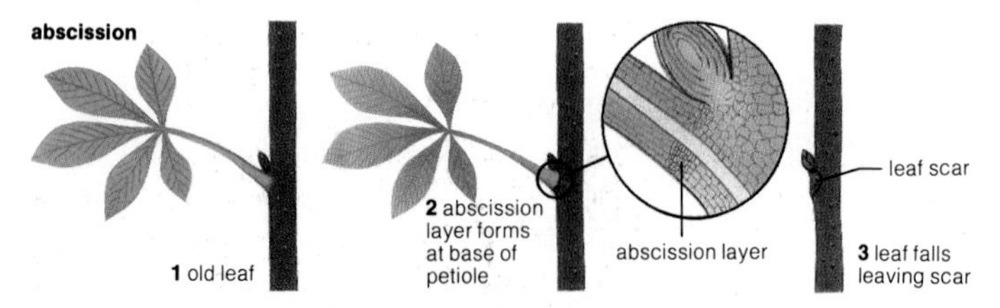

abscission (*n*) the process of separation of cells in the stalk of an organ (p. 88), e.g. the petiole (p. 96) of a leaf, leading to the dropping of the organ.

ethene (*n*) C_2H_4. A simple plant hormone (p. 112) which affects tropisms (↓), the inhibition (p. 14) of root growth, abscission (↑), the ripening (p. 83) of fruit, and other growth processes. Also known as **ethylene**.

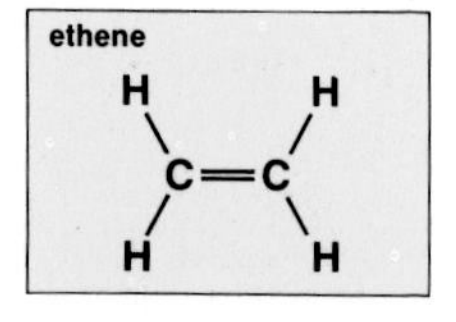

florigen (*n*) a possible hormone (p. 112) involved in the production of flowers.

synergistic (*adj*) of processes in which one substance reinforces the action of another substance or substances. This is usually applied to plant hormones (p. 112), which often affect each other and control similar growth processes. **synergism** (*n*).

climacteric (*n*) a period of high CO_2 output, controlled by the hormone (p. 112) ethene (↑), at the beginning of fruit ripening (p. 83).

apical dominance the inhibition (p. 14) of the development of lateral (p. 92) buds (p. 110) by hormones (p. 112) produced in the shoot apex (p. 90). If the apex is cut off, the buds develop.

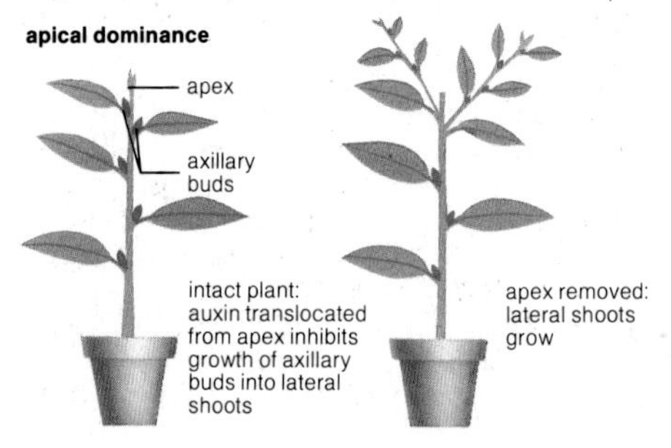

tropism (*n*) curving growth of a plant organ (p. 88) due to a stimulus (p. 170) coming from a particular direction, e.g. light or gravity.

phototropism

auxin in phototropism

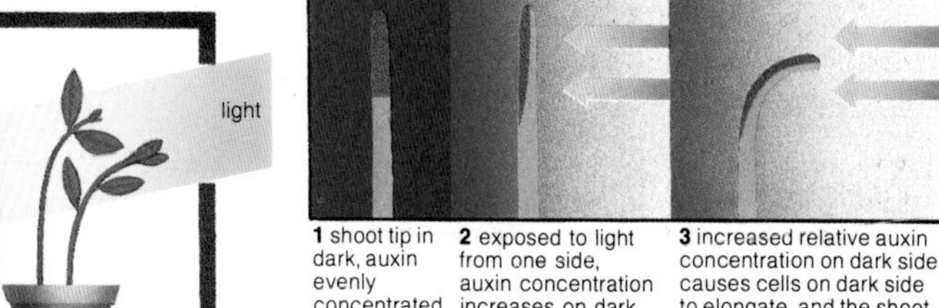

1 shoot tip in dark, auxin evenly concentrated

2 exposed to light from one side, auxin concentration increases on dark side and decreases on light side

3 increased relative auxin concentration on dark side causes cells on dark side to elongate, and the shoot bends towards the light

geotropism

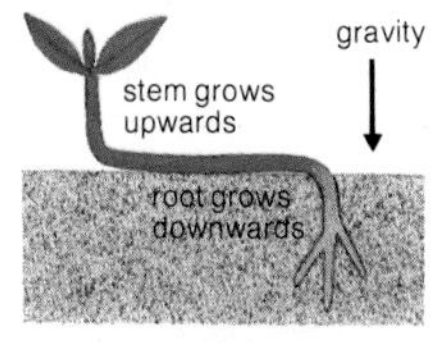

plagiogeotropism

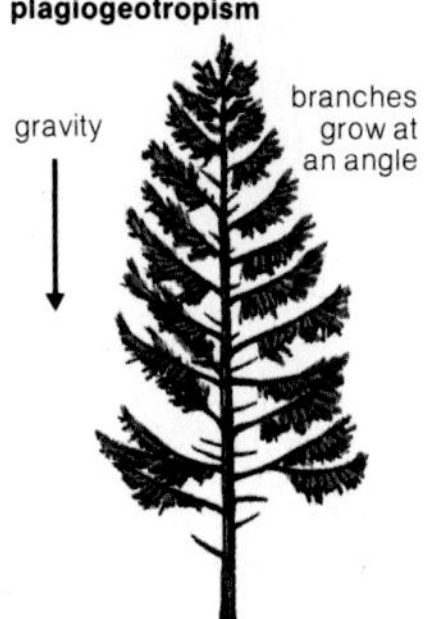

phototropism (*n*) curving growth of a plant organ (p. 88), e.g. a shoot, towards light coming from a particular direction. **phototropic** (*adj*).

geotropism (*n*) curving growth of a plant organ (p. 88) due to gravity. Geotropism can be downwards (positive), e.g. in a taproot (p. 88), or upwards (negative), e.g. in the shoot of a seedling (p. 87). **geotropic** (*adj*).

statolith (*n*) a very small grain of starch (p. 30), surrounded by a membrane (p. 18), often found in cells of growing tissues (p. 88). Statoliths are thought to be important in the control of geotropism (↑).

plagiogeotropism (*n*) growth at an angle, in response to gravity, e.g. in lateral (p. 92) branches. **plagiogeotropic** (*adj*).

thigmotropism (*n*) curving growth due to contact with an object, e.g. the coiling of tendrils (p. 136) of a climbing plant around a stake. This is sometimes known as haptotropism.

chemotropism (*n*) curving growth of a plant organ (p. 88) in response to a chemical stimulus (p. 170) or gradient (p. 24). **chemotropic** (*adj.*).

hydrotropism (*n*) curved growth of a plant organ (p. 88) in response to the stimulus (p. 170) of moisture. **hydrotropic** (*adj.*).

nastic movement any plant movement caused by a diffuse stimulus (p. 170), e.g. the changes in leaf position which occur in some plants at night.

endogenous rhythm the repeated, regular, rhythmic changes of internal activity in an organism that are not due to external environmental (p. 149) factors.

photoperiod (*n*) the number of hours of daylight needed by a plant before it will begin to flower. *See also* **photoperiodism** (↓).

short-day plant a plant which will begin to flower when the length of the day, or photoperiod (↑), is shorter than a certain period.

long-day plant a plant which will begin to flower only when the length of the day, or photoperiod (↑), is longer than a certain period.

photoperiodism (*n*) the physiological (p. 111) responses of an organism to changes in the lengths of day and night. **photoperiodic** (*adj*).

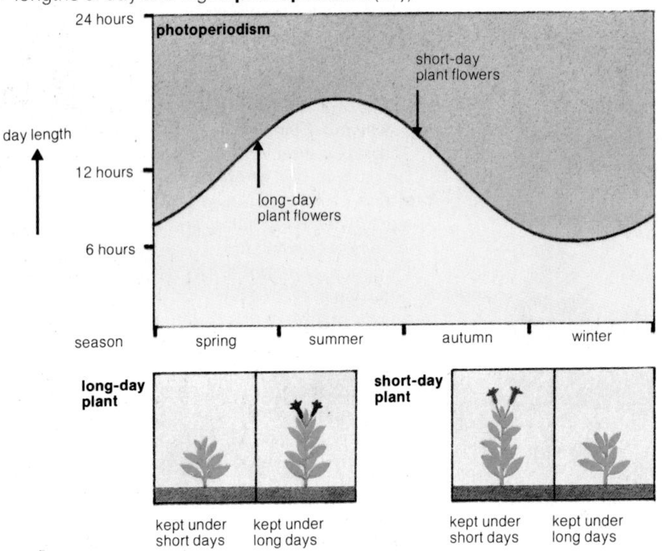

phytochrome (*n*) a pigment (p. 36) which controls many of the physiological (p. 111) responses of plants to light, e.g. photoperiodism (↑). Phytochrome absorbs the red and far-red wavelengths (p. 38) of light.

ephemeral (*adj*) of plants which germinate, grow, reproduce (p. 59) and die in a very short time.

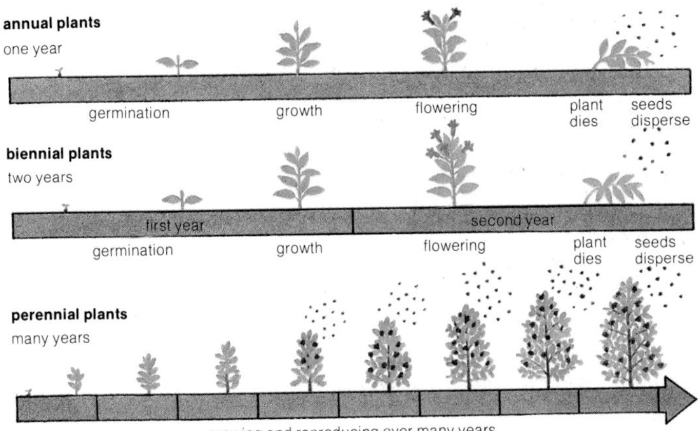

annual (*adj*) of plants which complete their entire life cycle (p. 64), from seed to reproduction (p. 59) to death, in one year.

biennial (*adj*) of plants which complete their life cycle (p. 64) in two years, growing in the first, reproducing (p. 59) and dying in the second.

perennial (*adj*) of plants which grow and reproduce (p. 59) for many years. Perennial plants are usually woody.

perennation (*n*) the survival of an individual (p. 135) over successive years, or of a dormant (↓) organ (p. 88) during unfavourable seasons. **perennate** (*v*).

dormant (*adj*) of cells, buds (p. 110), seeds, etc. during the period before growth begins. **dormancy** (*n*).

hibernation (*n*) the slowing-down of metabolism (p. 14) which occurs in many organisms during winter. **hibernate** (*v*).

vernalization (*n*) flowering due to treatment with low temperatures.

senescence (*n*) the process of growing old before death. **senescent** (*adj*).

organism (*n*) any living thing. Organisms differ from non-living things in that they can grow and reproduce (p. 59).

microorganism (*n*) a very small organism e.g. a virus (↓), bacterium (↓) or yeast (p. 164).

plant (*n*) an organism which has most or all of the following characteristics: ability to synthesize (p. 13) carbohydrate (p. 28) by photosynthesis (p. 32), possession of cell walls (p. 17) containing cellulose (p. 17), a life cycle (p. 64) consisting of an alternation of generations (p. 64), and an inability to move.

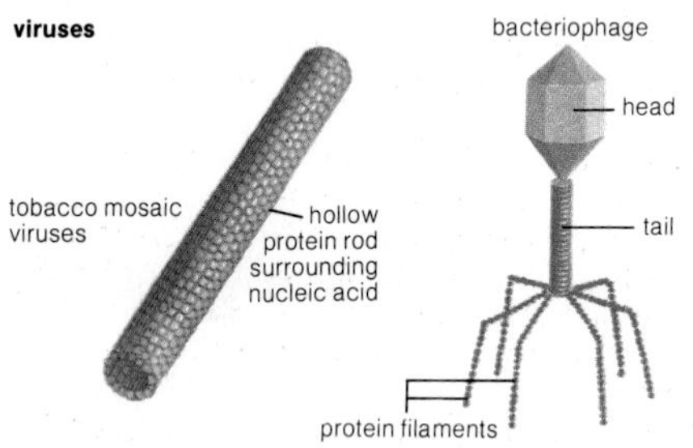

bacteriophage attacking a bacterium

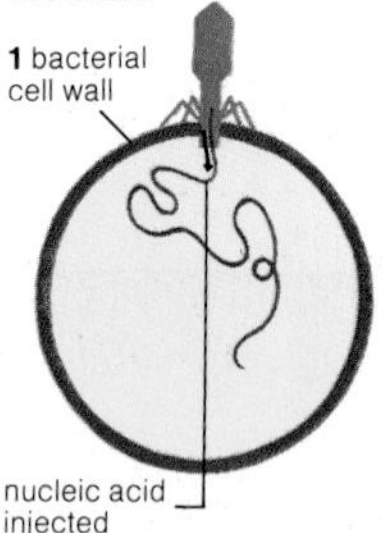

2 parts of new bacteriophages synthesized in bacterial cell

3 bacterium destroyed, new bacteriophages released

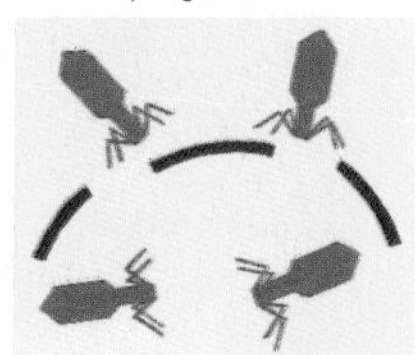

virus (*n*) a group of very simple organisms, which consist of a strand of nucleic acid (p. 51) surrounded by a coat of protein (p. 56). Viruses do not metabolize (p. 14), and can only reproduce (p. 59) inside the cells of other organisms, where the nucleic acid of the virus directs the synthesis (p. 13) of more viruses using the protein synthesis (p. 57) machinery already in the cell. Viruses can destroy cells in this way, and cause many diseases in other organisms. They are sometimes classified (p. 132) in a kingdom (p. 134) of their own. Viruses are very small, usually about 100 nm wide.

bacteriophage (*n*) a kind of virus (↑) which attacks the cells of bacteria (↓). Many bacteriophages have a 'head', consisting of a protein (p. 56) coat containing nucleic acid (p. 51), and a 'tail' of protein through which the nucleic acid is injected into a bacterial cell.

phage (*n*) = a bacteriophage (↑).

bacteria

bacilli

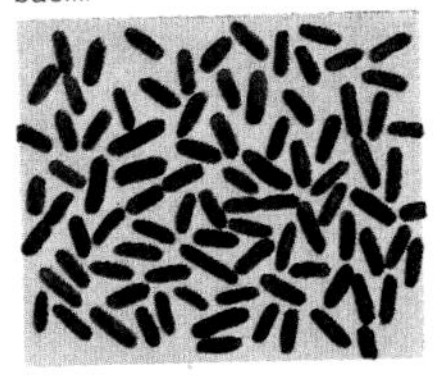

spirochaetes

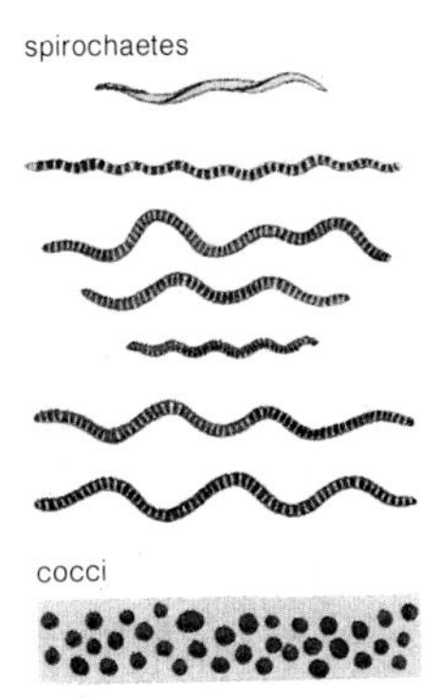

cocci

bacteria (*n.pl.*) a division (p. 134) of unicellular (↓) prokaryotic (p. 16) organisms, most of which are heterotrophic (p. 32). Bacterial cells are usually 0.5–2 μm across. Bacteria are important in the decay (p. 157) of organic (p. 11) matter in the soil. Many are parasitic (p. 144) on other organisms, often causing disease. **bacterium** (*sing.*), **bacterial** (*adj*).

bacillus (*n*) a rod-shaped bacterium (↑).

spirochaete (*n*) a spirally-shaped bacterium (↑), belonging to the order (p. 134) Spirochaetales.

cocci (*n*) a group of bacteria (↑) with spherically-shaped cells.

unicellular (*adj*) of organisms which consist of only one cell, e.g. euglenoids (p. 120), yeast (p. 164), bacteria (↑).

multicellular (*adj*) of organisms which consist of many cells, as in most plants.

algae (*n.pl.*) a large group of mainly aquatic (p. 161) plants, which differ from other plants by their lack of complex multicellular (↑) reproductive (p. 59) organs (p. 88). **alga** (*sing.*), **algal** (*adj*).

pyrenoid (*n*) a small grain of protein (p. 56) in the chloroplast (p. 32) of an algal (↑) cell, around which starch (p. 30) is deposited.

colony (*n*) a group of cells of the same kind, forming a single organism, as in many algae (↑). **colonial** (*adj*).

coenobium (*n*) a colony (↑) with a regular shape, consisting of cells which do not divide vegetatively (p. 60), e.g. the alga (↑) *Volvox*.

aggregation (*n*) a group of similar cells without a regular arrangement, as in many algae (↑).

colonial algae

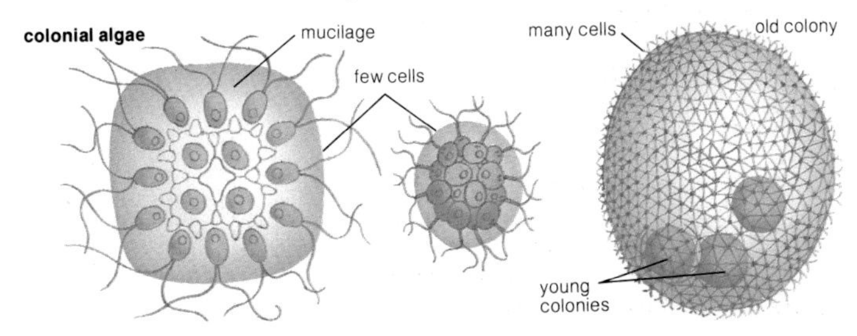

chromatophore (*n*) the chloroplast (p. 32) of a green alga (↓), or, the pigment (p. 36)-containing body of a photosynthetic (p. 32) bacterium (p. 119).

paramylum (*n*) a polysaccharide (p. 30) made of glucose (p. 28) units. It is stored in grains in euglenoids (↓).

siphoneous (*adj*) of algae (p. 119) in which the plant body is not divided into cells, i.e. it is multinucleate (p. 163).

parenchymatous (*adj*) of multicellular (p. 119) algae (p. 119) whose cells divide in more than one direction.

filamentous (*adj*) of algae (p. 119) consisting of long threads of cells, e.g. *Spirogyra*.

coccoid (*adj*) of algae (p. 119) which are unicellular (p. 119) and non-motile (↓).

filamentous blue green algae

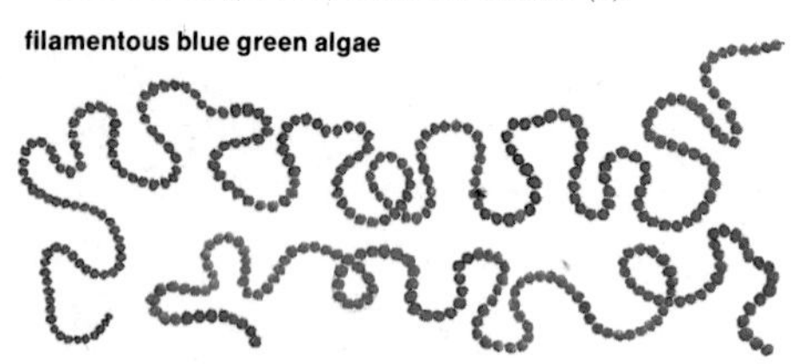

blue-green algae algae (p. 119) of the division (p. 134) Cyanophyta with prokaryotic (p. 16) cells. They are unicellular (p. 119) or multicellular (p. 119), lack flagella (↓) and have their own carotenoid (p. 37) photosynthetic (p. 32) pigments (p. 36).

red algae algae (p. 119) of the division (p. 134) Rhodophyta which have a red colour due to the presence of the pigments (p. 36) phycocyanin and phycoerythrin.

green algae algae (p. 119) of the division (p. 134) Chlorophyta. Green algae have chlorophyll *b* (p. 36), produce starch (p. 30), and have cellulose (p. 17) in their cell walls (p. 17).

euglenoid (*n*) an alga (p. 119) of the division (p. 134) Euglenophyta. Euglenoids are unicellular (p. 119), flagellate (↓) and have paramylum (↑) instead of starch (p. 30) as their main storage product.

euglenoid

Euglena

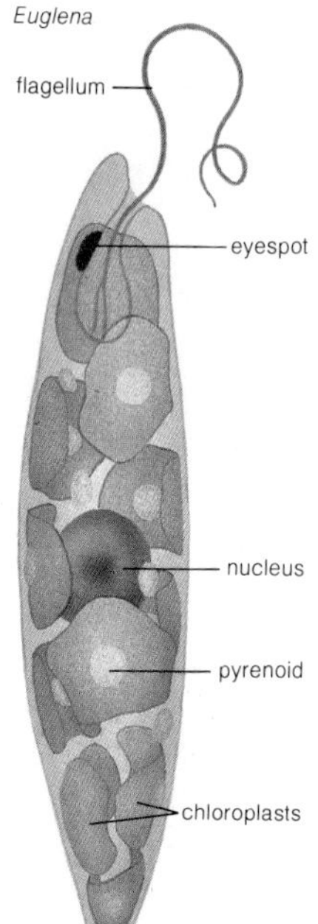

flagellum (*n*) a long motile (↓) thread, consisting of a membrane (p. 18) enclosing a series of parallel (p. 110) microtubules (p. 21). Flagella (*pl.*) are found in the cells of unicellular (p. 119) motile algae (p. 119), e.g. *Euglena*, *Chlamydomonas*, and also in the male gametes (p. 61) of bryophytes (p. 122), pteridophytes (p. 126), and some gymnosperms (p. 128). **flagellate** (*adj*).

motile (*adj*) able to move, as in cells with flagella (↑).

diatoms

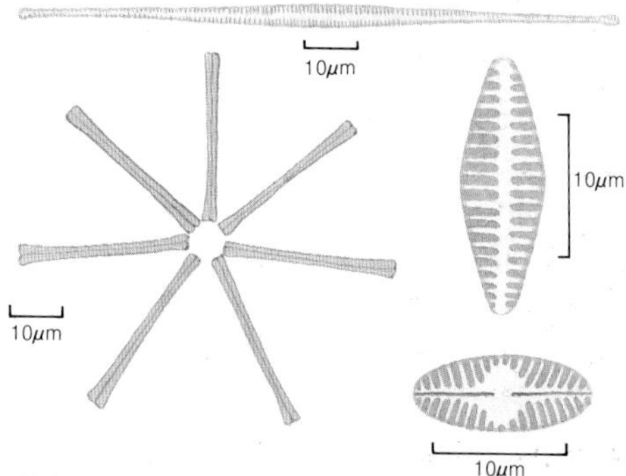

diatom (*n*) an alga (p. 119) of the division (p. 134) Bacillariophyta. Diatoms are mostly unicellular (p. 119), and their cell walls (p. 17) contain silicon. *See also* **siliceous skeleton**.

siliceous skeleton the silicon-containing cell wall (p. 17) of a diatom (↑).

dinoflagellate

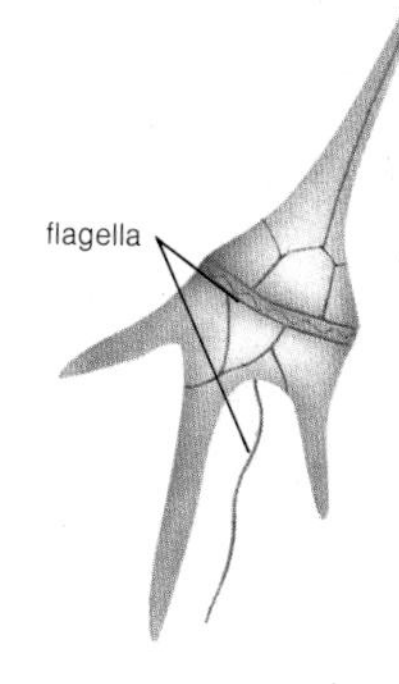

dinoflagellate (*n*) a class (p. 134) of unicellular (p. 119) algae (p. 119), usually yellow in colour, which have two flagella (↑) and thick cell walls (p. 17) arranged in a characteristic pattern of plates. Dinoflagellates are a major component of marine phytoplankton (↓).

phytoplankton (*n*) the small plants, mostly diatoms (↑) and other unicellular (p. 119) algae (p. 119), which are found near the surface of oceans and lakes. Phytoplankton is one of the world's most important primary producers (p. 150).

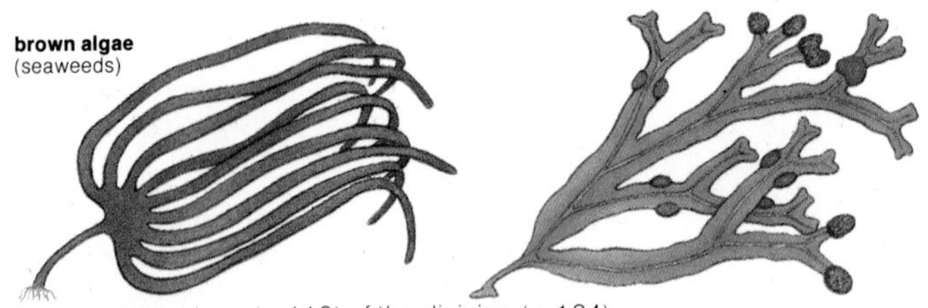

brown algae algae (p. 119) of the division (p. 134) Phaeophyceae, including many of the larger seaweeds (↓). Brown algae contain brown accessory pigments (p. 36).

seaweed (*n*) the general name for any large, parenchymatous (p. 120) alga (p. 119) in the sea.

nonvascular (*adj*) of plants which do not have a vascular system (p. 105). Nonvascular plants include most bryophytes (↓) and all algae (p. 119).

vascular (*adj*) of plants with a vascular system (p. 105). Vascular plants include all pteridophytes (p. 126) and spermatophytes (p. 128).

bryophyte (*n*) a plant of the division (p. 134) Bryophyta, i.e. a moss (p. 124) or a liverwort (↓). Bryophytes differ from other plants in having the gametophyte (p. 65) as the main vegetative (p. 60) stage. Most bryophytes have little or no vascular system (p. 105), and live in wet, shady habitats (p. 149).

thallus (*n*) a more or less undifferentiated (p. 110) plant body, without distinct roots, stems and leaves, e.g. the gametophyte (p. 65) of thalloid liverworts (↓), or the plant body of an alga (p. 119). **thalloid** (*adj*), **thalli** (*pl.*).

prothallus (*n*) the gametophyte (p. 65) of mosses (p. 124), liverworts (↓) and pteridophytes (p. 126). **prothalli** (*pl.*).

rhizoid (*n*) a thread-like cell which grows from the lower surface or base of a bryophyte (↑). Rhizoids have the function of roots.

sporogonium (*n*) the sporophyte (p. 65) of a moss (p. 124) or a liverwort (↓), consisting of a foot (↓), seta (↓) and capsule (↓).

foot (*n*) the base of the sporophyte (p. 65) of a bryophyte (↑), which is the part that attaches it to the gametophyte (p. 65).

rhizoids

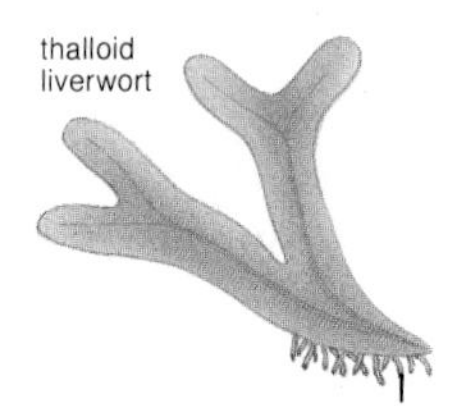

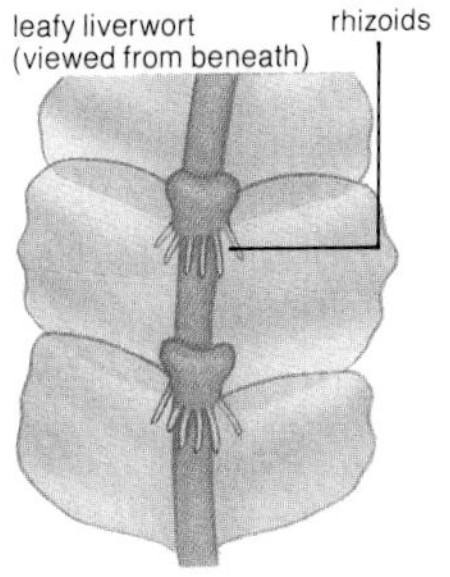

seta (*n*) the stalk of the sporophyte (p. 65) of a bryophyte (↑).

capsule[2] (*n*) the spore (p. 66)-producing organ (p. 88) of the sporophyte (p. 65) of a bryophyte (↑), borne at the top of the seta (↑).

liverwort (*n*) one of the two groups of bryophytes (↑). Liverworts differ from mosses (p. 124) in having less differentiated (p. 110) cells in the gametophyte (p. 65), and in having elaters (p. 124) in the dehiscent (p. 84) capsule (↑). The gametophyte is either thalloid (↑) or leafy.

hepatic (*n*) = a liverwort (↑).

liverwort sporophyte

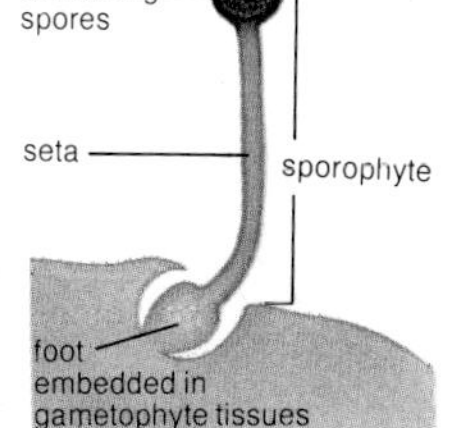

thalloid liverwort

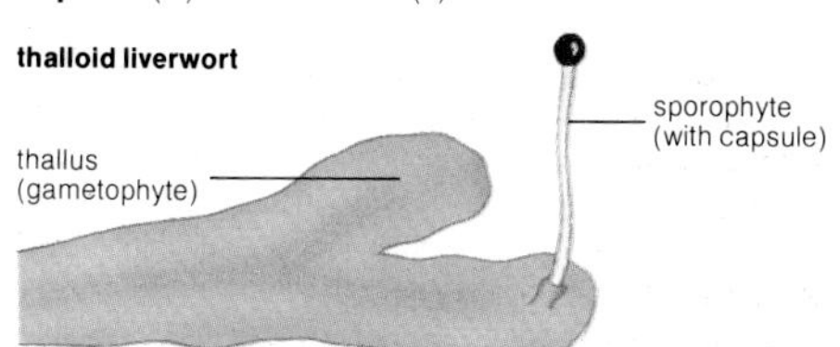

thalloid liverwort a liverwort (↑) in which the gametophyte (p. 65) is a flat, more or less undifferentiated (p. 110) thallus (↑). About 20% of liverwort species (p. 134) are thalloid.

leafy liverworts
(viewed from above)

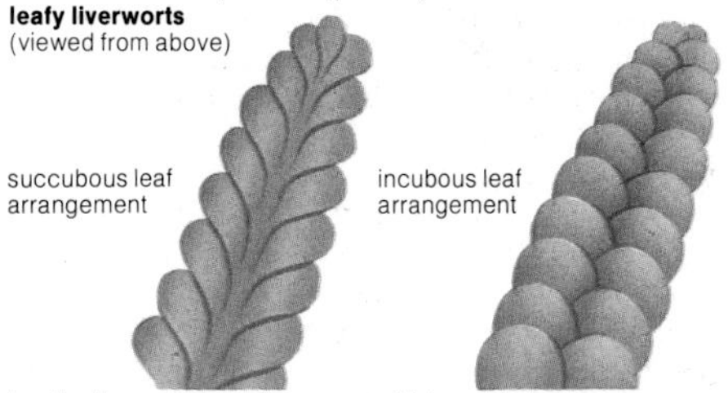

leafy liverwort a liverwort (↑) in which the gametophyte (p. 65) has a simple stem, growing from the apex (p. 90) and bearing small leaves in rows along it. About 80% of liverwort species (p. 134) are leafy.

succubous (*adj*) of a kind of growth in leafy liverworts (↑), in which the front of each leaf lies underneath the leaf in front of it.

incubous (*adj*) of a kind of growth in leafy liverworts (↑), in which the front of each leaf lies on top of the leaf in front of it.

gemmae in thalloid liverworts: vegetative reproduction

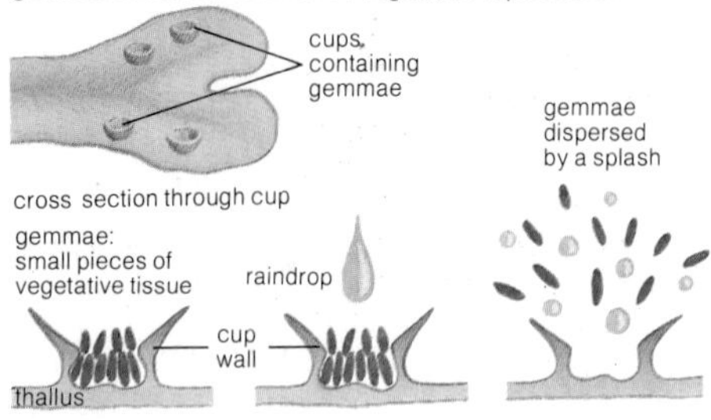

liverwort sporophyte discharging spores

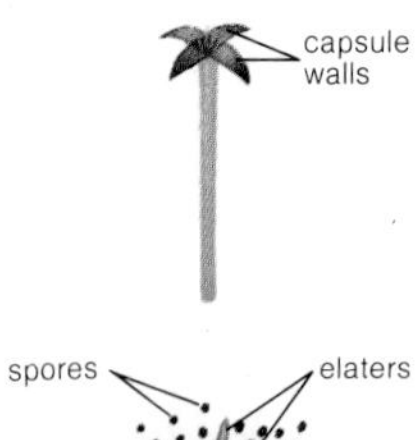

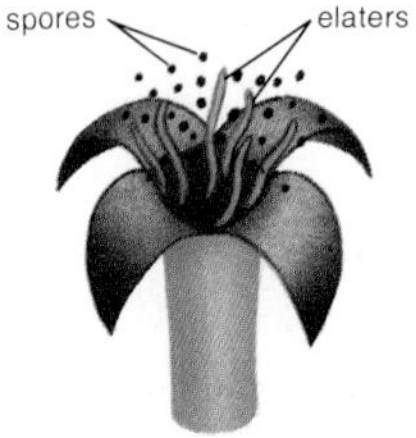

elater with helical thickenings in cell wall

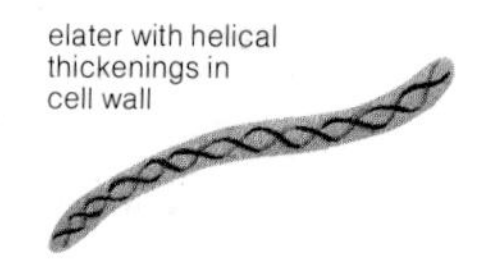

gemmae (*n.pl.*) small groups of green cells produced in cup-shaped structures on the surface of some thalloid liverworts (p. 123). Gemmae are dispersed (p. 84) by splashes of rain, and are a way of vegetative reproduction (p. 60). **gemma** (*sing.*).

elater (*n*) one of a bunch of long, thin cells in the capsule (p. 123) of the sporophyte (p. 65) of a liverwort (p. 123). Elaters have spiral (p. 98) thickening of the cell wall (p. 17). They alter their position with changes in humidity, and help with the dispersal (p. 84) of spores (p. 66) from the capsule.

moss (*n*) one of the two main groups of bryophytes (p. 122). Mosses differ from liverworts (p. 123) in having more differentiated (p. 110) cells in the gametophyte (p. 65). The gametophyte usually has a stem with leaves, and is often branched. The capsule (p. 123) of the sporophyte (p. 65) is also more differentiated in mosses, and the spores (p. 66) are released through a peristome (↓).

moss

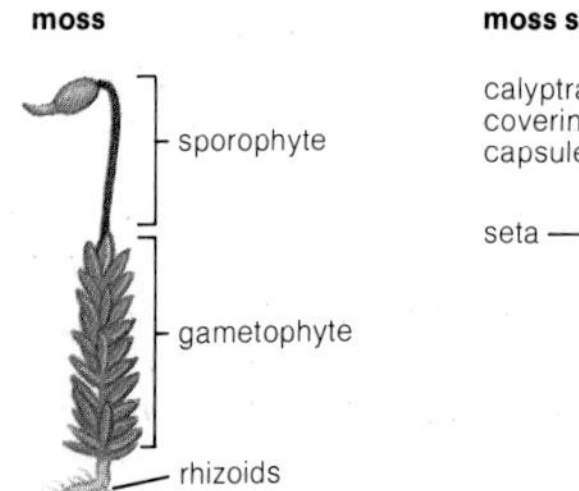

moss sporophyte

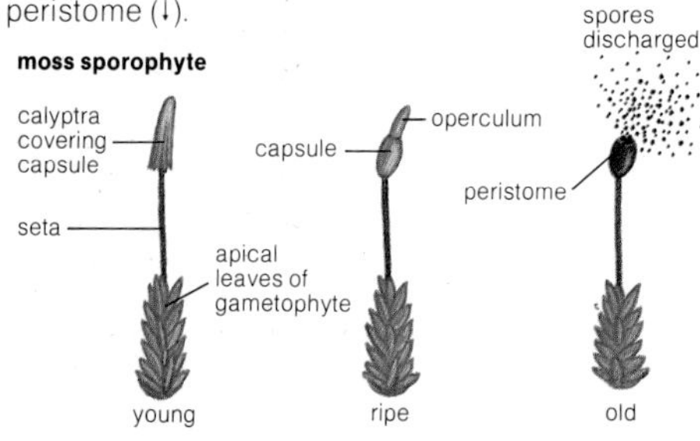

protonema

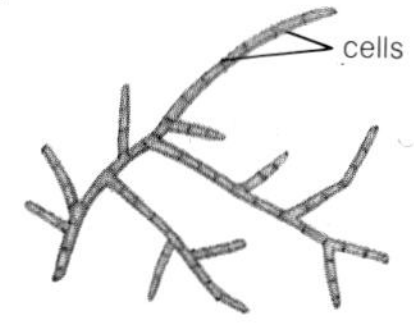

acrocarpous moss

pleurocarpous moss

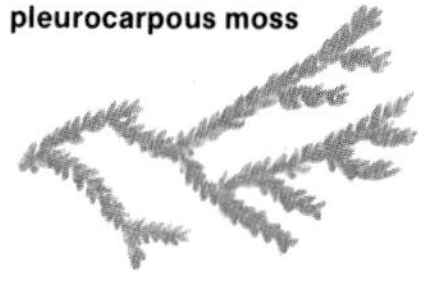

protonema (*n*) the young gametophyte (p. 65) of a moss (↑), in the early stages after the germination (p. 87) of the spore (p. 66).

pleurocarpous (*adj*) of mosses (↑) which have a stem with many branches, spreading across the ground. The reproductive (p. 59) organs (p. 88) are borne on short side-branches.

acrocarpous (*adj*) of mosses (↑) which have an upright stem, with the reproductive (p. 59) organs (↑) at the apex (p. 90).

paraphyses (*n.pl.*) small hairs one cell thick, often with a large round cell at the top, which grow among the antheridia (p. 65) in mosses (↑). They protect the antheridia and may provide them with photosynthetic (p. 32) products.

paraphyses

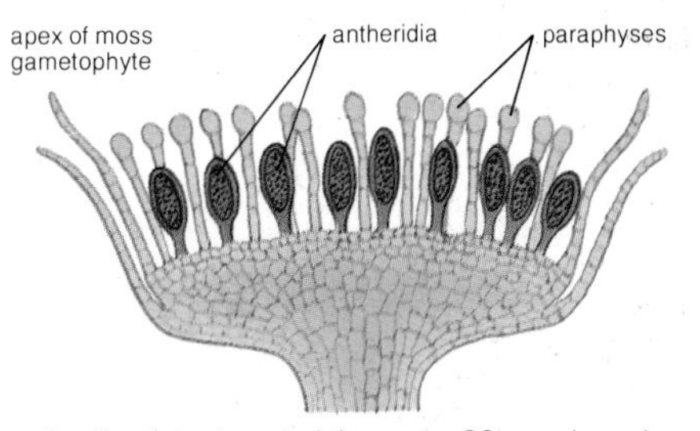

calyptra (*n*) a hood of tissue (p. 88) produced from the wall of the archegonium (p. 65), especially in mosses (↑). The calyptra protects the young sporophyte (p. 65).

columella[1] (*n*) the tissue (p. 88) in the centre of the capsule (p. 123) of a moss (↑).

operculum (*n*) the lid which covers the pore (p. 19) at the apex (p. 90) of a moss (↑) capsule (p. 123). The operculum opens to allow spores (p. 66) to escape.

peristome (*n*) a set of tooth-like plates under the operculum (↑) of the capsule (p. 123) of a moss (↑). These plates are called peristome teeth. They control the release of spores (p. 66) into the air, by altering their position with changes in humidity.

pteridophyte (*n*) a member of the division (p. 134) Pteridophyta, which includes the ferns (↓), clubmosses (↓) and horsetails (↓). In pteridophytes the sporophyte (p. 65) is the main vegetative (p. 60) stage of the life cycle (p. 64). The gametophyte (p. 65) is small, and is independent of the sporophyte, as in bryophytes (p. 122). Large tree-like forms of pteridophytes were common during the Carboniferous (p. 143) period, and were fossilized (p. 142) to form coal.

fern (*n*) a pteridophyte (↑) belonging to the order (p. 134) Filicales. Ferns have spirally (p. 98) arranged leaves, which are often pinnately (p. 98) compound (p. 98). Ferns are homosporous (p. 67), with the sporangia (p. 66) borne in sori (↓) on the abaxial (p. 97) surface of the leaf.

frond (*n*) the leaf of a fern (↑). Most ferns have pinnate (p. 98) or bipinnate (p. 98) fronds. The leaves of palms (p. 130) are also called fronds.

circinate (*adj*) coiled or rolled up, like the young frond (↑) of a fern (↑).

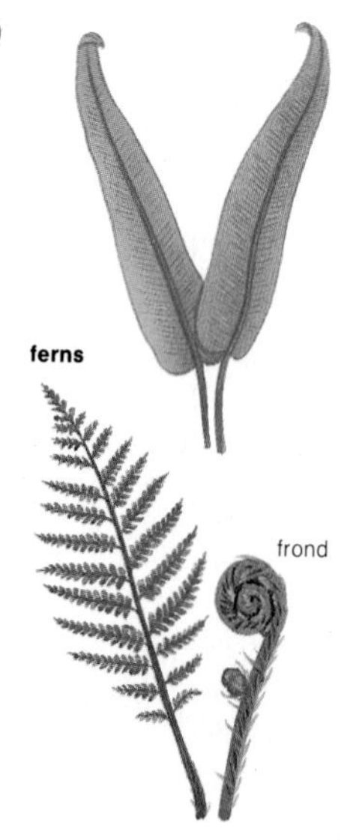

types of sori on underside of fern leaves

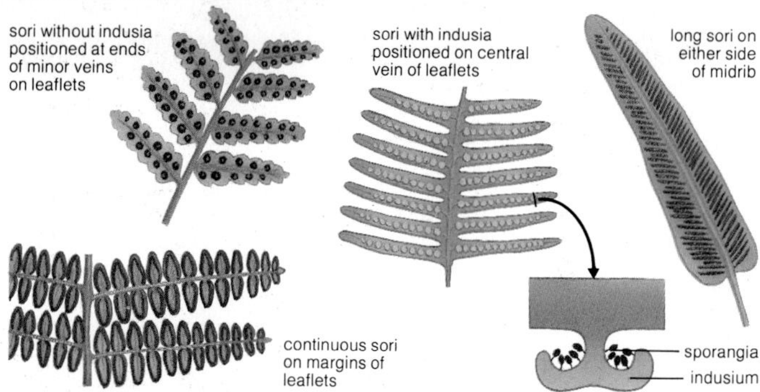

sorus (*n*) an organ (p. 88) on the surface of the leaf of a fern (↑), in which sporangia (p. 66) are produced. Sori (*pl.*) have different shapes and are found on different parts of the leaf in different species (p. 134). Their function is to protect the sporangia.

indusium (*n*) the flap of tissue (p. 88) in a sorus (↑), which protects the sporangia (p. 66). **indusia** (*pl.*).

filmy fern a fern of the family (p. 134) Hymenophyllaceae. Filmy ferns have very delicate leaves, usually only one cell thick, and live in moist shady habitats (p. 149).

tree fern a fern (↑) of the family (p. 134) Cyatheaceae, in which the fronds (↑) grow from the top of a trunk (p. 92). The trunk in tree ferns consists partly of the woody bases of dead fronds. Most tree ferns are tropical (p. 162).

tree fern

horsetail (*Equisetum*)

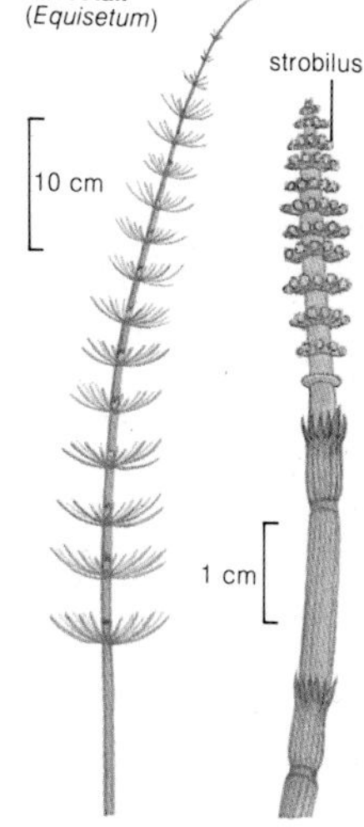

clubmoss (*n*) a pteridophyte (↑) of the order (p. 134) Lycopodiales, e.g. *Lycopodium*. Clubmosses are not related to mosses (p. 124).

horsetail (*n*) a pteridophyte (↑) of the order (p. 134) Equisetales, which consists of about 25 species (p. 134) of *Equisetum*.

clubmoss (*Lycopodium*)

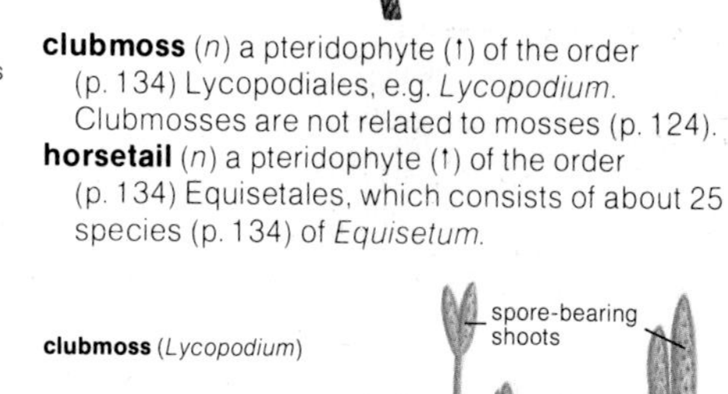

cryptogam (*n*) a general name for all plants except gymnosperms (↓) and angiosperms (p. 130), in an old classification (p. 132) of the plant kingdom (p. 134). Cryptogams reproduce (p. 59) by spores (p. 66).

phanerogam (*n*) a general name for all seed plants (↓) in an old classification (p. 132) of the plant kingdom (p. 134), so called because the organs (p. 88) of reproduction (p. 59) are clearly visible. This name has been replaced by spermatophyte (↓).

spermatophyte (*n*) a member of the division (p. 134) Spermatophyta, or seed plants (↓). This division includes all angiosperms (p. 130) and gymnosperms (↓).

seed plant a plant which reproduces (p. 59) by seed, i.e. a spermatophyte (↑) or phanerogam (↑).

gymnosperm (*n*) a spermatophyte (↑) of the subdivision (p. 134) Gymnospermae. Gymnosperms differ from angiosperms (p. 130) in their unprotected ovules (p. 78), arrangement of reproductive (p. 59) organs (p. 88) in cones (p. 68), archegonia (p. 65) and wood without vessels (p. 107).

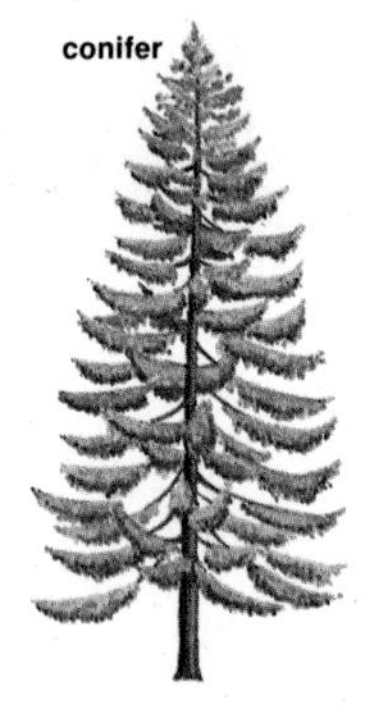

conifer (*n*) a gymnosperm (↑) of the order (p. 134) Coniferales, which includes pines, yews, cedars, redwoods, etc. Most conifers are monoecious (p. 79), with separate male and female cones (p. 68). They are mostly evergreen (p. 136), with narrow pointed leaves. **coniferous** (*adj.*).

cycad (*n*) a gymnosperm (↑) of the order (p. 134) Cycadales. Cycads are dioecious (p. 79), with motile (p. 121) male gametes (p. 61). They have palm- or fern-like leaves and are found mainly in tropical (p. 162) habitats (p. 149). The cycads are a primitive (p. 141) group, and there are many fossils (p. 142) dating from the Mesozoic (p. 143).

Ginkgoales (*n*) an order (p. 134) of gymnosperms (↑) with only one living species, *Ginkgo biloba* (maidenhair tree), which is found in China. They have motile (p. 121) male gametes, like cycads (↑), but deciduous (p. 136) leaves.

Gnetales (*n*) a small order (p. 134) of gymnosperms (↑) made up of three genera (p. 134), *Ephedra*, *Gnetum*, and *Welwitschia*. Gnetales are similar to angiosperms (p. 130), because their wood contains vessels (p. 107) and their ovules (p. 78) do not have archegonia (p. 65).

angiosperm (*n*) a spermatophyte (p. 128) of the subdivision (p. 134) Angiospermae. They differ from gymnosperms (p. 128) in that their ovules (p. 78) are protected in the ovary (p. 76) and in their possession of vessels (p. 107) in the xylem (p. 106). They also have double fertilization (p. 78) of the ovum (p. 61) and the endosperm (p. 86). There are more than 200 families (p. 134) and 250 000 species (p. 134) of angiosperms. They are divided into two classes (p. 134), the Monocotyledones (↓) and the larger Dicotyledones (↓).

flowering plant a plant which bears flowers. This term is usually used only for angiosperms (↑), but is sometimes used for some gymnosperms (p. 128) as well.

monocotyledon (*n*) an angiosperm (↑) of the class (p. 134) Monocotyledones. Their seeds have one cotyledon (p. 86). Monocotyledons do not have secondary thickening (p. 94) and most are small herbaceous (p. 136) plants with parallel (p. 110) venation (p. 97) and floral parts in whorls (p. 98) of three or multiples of three, e.g. grasses (↓), sedges (↓), orchids (↓), palms (↓). **monocot** (*abbr.*).

grass (*n*) a monocotyledon (↑) of the family (p. 134) Gramineae (sometimes called Poaceae). It is a very large family, and contains all the cultivated cereals (wheat, rice, etc.).

sedge (*n*) a monocotyledon (↑) of the genus (p. 134) *Carex* in the family (p. 134) Cyperaceae.

orchid (*n*) a monocotyledon (↑) of the family (p. 134) Orchidaceae. Most orchids are tropical (p. 162), and epiphytic (p. 137). The family is one of the largest in the plant kingdom (p. 134), with at least 17 000 species (p. 134).

palm (*n*) a monocotyledon (↑) of the family (p. 134) Palmae. Palms are the largest monocotyledons, and are found mostly in tropical (p. 162) forests (p. 158). They usually have a thick unbranched pachycaul (p. 94) trunk (p. 92), with rings on the surface. These rings mark the position of fallen leaves. Most palms have compound (p. 98) leaves, borne in a thick crown (p. 92) at the top of the trunk.

monocotyledons

dicotyledon (*n*) an angiosperm (↑) of the class (p. 134) Dicotyledones. The seeds of dicotyledons have two cotyledons (p. 86). Dicotyledons have secondary thickening (p. 94) in the stems. Most families (p. 134) and species (p. 134) of angiosperms are dicotyledons. **dicot** (*abbr.*).

cactus (*n*) a dicotyledon (↑) of the family (p. 134) Cactaceae, mainly found in hot dry areas in North and South America. Cacti (*pl.*) usually have thick succulent (p. 99) stems and groups of spines (p. 100) instead of leaves.

Leguminosae (*n*) a large family (p. 134) of dicotyledons (↑). The seeds of Leguminosae are borne in legumes (p. 84) or pods (p. 84). Many species (p. 134) of Leguminosae are important crop plants, e.g. beans, peas, clover, etc.

Compositae (*n*) a large family (p. 134) of dicotyledons (↑). Compositae have composite (p. 81) inflorescences (p. 80), as in daisies.

classification (*n*) the naming of species (p. 134) and their grouping into families (p. 134), orders (p. 134), divisions (p. 134), etc. **classify** (*v*).

how a species is classified in the plant kingdom

common oak	*Quercus robur*	species
oaks	*Quercus*	genus
beeches, chestnuts, oaks	Fagaceae	family
beeches, chestnuts, oaks, birches, alders, hazels, hornbeams	Fagales	order
dicotyledons	Dicotyledones	class
flowering plants	Angiospermae	subdivision
seed plants	Spermatophyta	division
plants	Plantae	kingdom
common name	**Latin name**	**taxon**

nomenclature (*n*) the part of classification (↑) that involves the naming of species (p. 134), families (p. 134), orders (p. 134), etc.

systematics (*n*) the part of classification (↑) that involves the arrangement of organisms (p. 118) into related groups.

Linnaeus Carl Linnaeus, (1707-1778) was responsible for the modern system of naming plants and animals. This is the binomial (↓) system. His most famous work was *Species Plantarum*, published in 1753, in which he described all the plants then known to man.

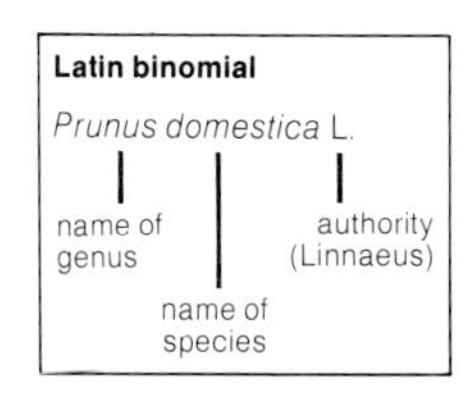

binomial (*n*) the Latin name of a species (p. 134), consisting of two words. The first is the name of the genus (p. 134) to which the species belongs, and the second is the name which distinguishes the species from other species in the same genus. This system of naming species was introduced by Linnaeus (↑). **binomial** (*adj*).

authority (*n*) the name of the author who was the first to give a species (p. 134) or other taxon (↓) its name. In the case of species, the authority is given after the binomial (↑).

herbarium (*n*) a collection of dried pressed plants, used by taxonomists in the classification (↑) of plants.

type (*n*) the specimen of the individual (p. 135) plant from which a species (p. 134) was first described.

artificial key a way of identifying plants by steps. Each step involves a choice between at least two different characters, each of which leads to another choice of two characters, finally leading to the right identification.

taxonomy (*n*) the science of classification (↑) and the relationships of organisms. **taxonomic** (*adj*), **taxonomist** (*n*).

taxon (*n*) any taxonomic (↑) group, e.g. a species (p. 134), a family (p. 134). All the members of a taxon share similar characteristics, which are different from those of other groups. **taxa** (*pl.*).

characteristic (*adj*) of characters by which an organism or group of organisms can be recognized. For example, flowers are characteristic of angiosperms (p, 130), wood is characteristic of trees. **characteristic** (*n*).

character (*n*) any part or shape of an organism that makes it possible to classify (↑) the organism. Characters used in classification include the arrangement of the reproductive (p. 59) organs (p. 88), the shape of leaves, etc.

kingdom (*n*) the largest of all the taxa (p. 133). In older systems of classification (p. 132), there are only two kingdoms - plants and animals. In some newer systems, there are five - plants, animals, fungi (p. 163), bacteria (p. 119) and viruses (p. 118); before, the fungi, bacteria and viruses were included with the plants.
division (*n*) a major taxon (p. 133), e.g. bryophytes, which is made up of classes (↓). There are three divisions of land plants; these are bryophytes (p. 122), pteridophytes (p. 126) and spermatophytes (p. 128).
subdivision (*n*) a taxon (p. 133) within a division (↑).
class (*n*) a taxon (p. 133) consisting of orders (↓). Dicotyledons (p. 131) are a class.
order (*n*) a taxon (p. 133) consisting of families (↓). The Latin names of orders usually end with -ales.
family (*n*) a taxon (p. 133) consisting of related genera (↓). The Latin names of families usually end with -aceae.
tribe (*n*) a group of related genera (↓) within a family (↑).
genus (*n*) a group of related species (↓). The name of the genus is the first in a Latin binomial (p. 133). **genera** (*pl.*), **generic** (*adj*).
monotypic (*adj*) of a genus (↑) with only one species (↓), or a family (↑) with only one genus.
species (*n*) usually, the smallest unit of classification (p. 132). A species includes individuals (↓) which are alike and can breed (p. 59) with each other. Species are given Latin binomial (p. 133) names. They are sometimes divided into subspecies (↓) and varieties (↓) on the basis of small differences between populations (↓). **specific** (*adj*).

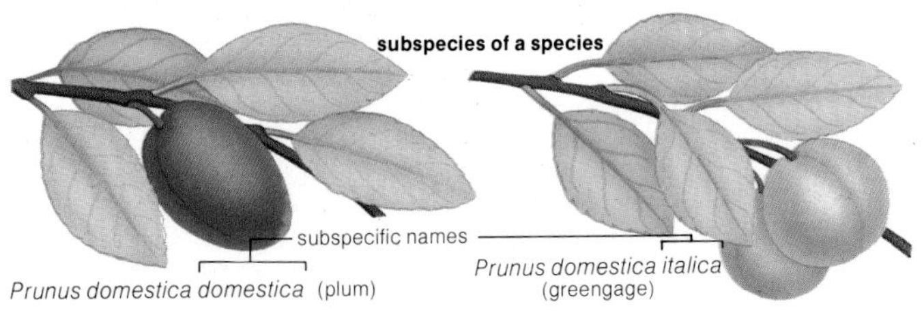

subspecies (*n*) a taxon (p. 133) within a species (↑). Subspecies of a species differ in small ways. Although they can breed (p. 59) with each other, they are usually found in different places, or different populations (↓). In the naming of subspecies, a third Latin name is put after the binomial (p. 133). **subspecific** (*adj*).

variety (*n*) a taxonomic (p. 133) group within a species (↑) or a subspecies (↑). The differences between varieties are small, and do not necessarily relate to differences in habitat (p. 149) or place.

strain (*n*) a reproductively (p. 59) isolated (p. 142) population (↓) whose individuals (↓) have identical genotypes (p. 41) over many generations (p. 63) and show phenotypic (p. 41) differences from other populations.

ecotype (*n*) a set of individuals (↓) or populations (↓) in a particular habitat (p. 149) that differ phenotypically from members of the same species in other habitats.

cline (*n*) continuous or gradual variation (↓), within a population (↓), in some of the characters of a species (↑). This variation is related to gradual changes in ecological (p. 149) conditions that occur, for instance, up the side of a mountain.

individual (*n*) a single organism.

variation (*n*) differences in the characteristics of individuals (↑) of a species (↑) or a population (↓).

infraspecific (*adj*) of variation (↑) between individuals (↑) of a species (↑).

polymorphism (*n*) the occurrence of two or more forms of a species (↑) in the same population (↓) or habitat (p. 149). **polymorphic** (*adj*).

morph (*n*) one of the forms of a polymorphic (↑) species (↑).

population (*n*) a group of individuals (↑) of the same species (↑) living in the same place or area, close enough to breed (p. 59) together.

endemic (*adj*) of taxa (p. 133) that are found only in one particular place or area. **endemism** (*n*).

distribution (*n*) the whole geographical range in which a taxon (p. 133) is found.

flora (*n*) the total of plant species (↑) in a particular region, country, continent, etc.

habit (*n*) the appearance of a plant, e.g. herb (↓), shrub (↓), tree (↓).
tree (*n*) a large, perennial (p. 117) plant with a woody trunk (p. 92), which usually bears branches.
sapling (*n*) a young tree (↑).
evergreen (*adj*) having green leaves throughout the year. **evergreen** (*n*).
deciduous (*adj*) (1) of plants which shed their leaves at least once a year and remain leafless for weeks or months; (2) of short-lived organs (p. 88) that are shed from a plant.
shrub (*n*) a small woody perennial (p. 117) plant, with branches from ground level upwards.

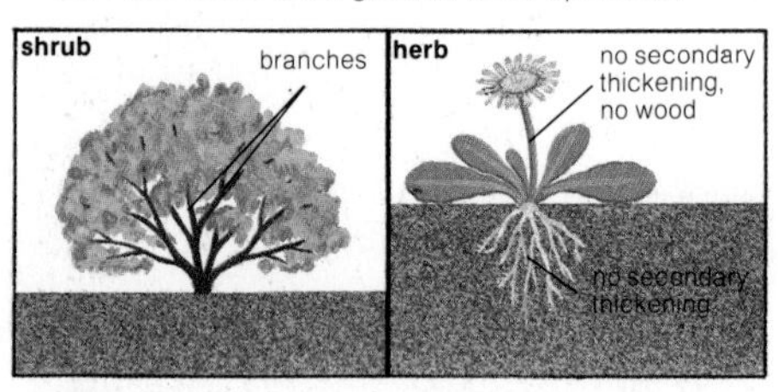

bush (*n*) a shrub (↑) with especially dense branching and foliage (p. 96).
herb (*n*) a small plant without wood in the stems or roots. **herbaceous** (*adj*).
climber (*n*) a plant with roots in the ground, which uses other plants to support itself. Climbers use tendrils (↓), adventitious roots (p. 89), or sucker-like discs to hold on to other plants, or sometimes twist around their stems.
tendril (*n*) a long, coiled, threadlike organ (p. 88), borne on the stems or leaves of many climbers (↑). Tendrils coil around the stems and branches of other plants, and help the climber to support itself. Tendrils are modified leaves, stems, leaflets (p. 98) or stipules (p. 99).
creeper (*n*) a plant which cannot support itself and spreads along the ground. Creepers usually have little or no secondary thickening (p. 94).
vine (*n*) (1) a climber (↑); (2) a member of the family (p. 134) Vitaceae.

two types of tree

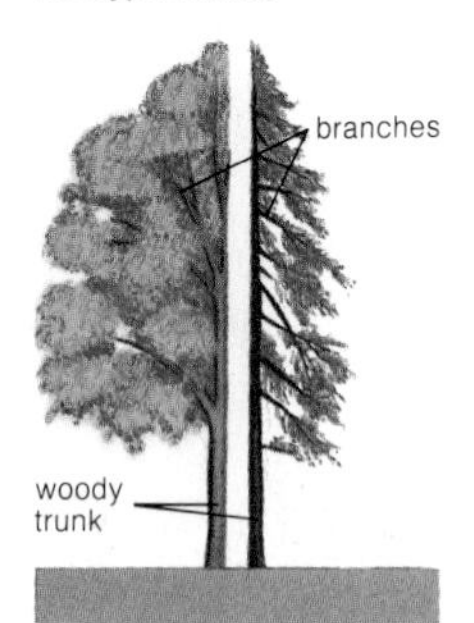

climbers
three examples of climbing plants

1 *Lathyrus* a plant that climbs by means of tendrils

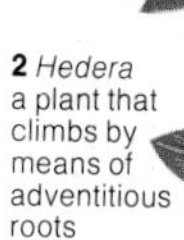

2 *Hedera* a plant that climbs by means of adventitious roots

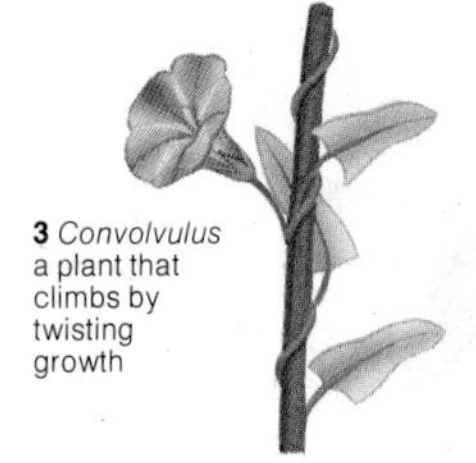

3 *Convolvulus* a plant that climbs by twisting growth

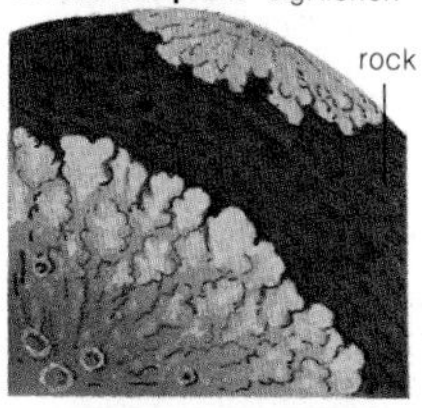

liana (*n*) a woody perennial (p. 117) climber (↑). Lianas are very common in tropical (p. 162) rain forests (p. 158).

saprophyte (*n*) a plant which obtains all its nutrients (p. 111) from dead organic (p. 11) matter. Saprophytes are heterotrophic (p. 32). Many fungi (p. 163), bacteria (p. 119) and some vascular (p. 122) plants are saprophytes in the soil. They are important in the cycles of inorganic (p. 11) nutrients in ecosystems (p. 149). **saprophytic** (*adj*).

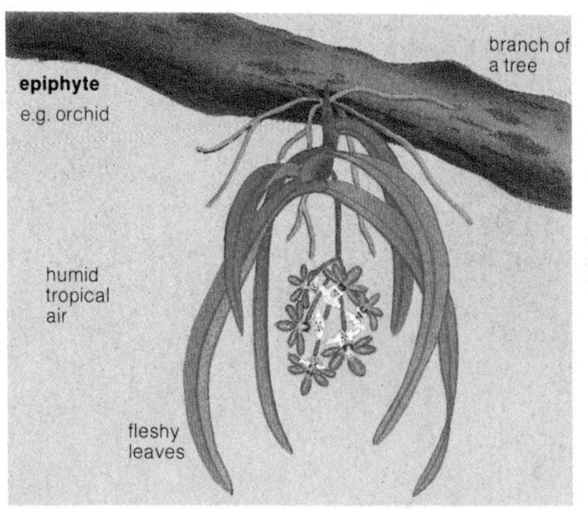

epiphyte (*n*) a plant which grows on the stems and branches of other plants. Epiphytes have no roots in the ground. They are not parasitic (p. 144). **epiphytic** (*adj*).

epiphyll (*n*) a plant which grows on the leaves of other plants. Most epiphylls are tropical (p. 162) liverworts (p. 123) and lichens (p. 147). **epiphyllous** (*adj*).

saxicolous (*adj*) of plants that live on rocks, e.g. lichens (p. 147).

xerophyte (*n*) a plant that lives in a desert or other dry habitat (p. 149). **xerophytic** (*adj*).

xeromorphic (*adj*) of plants with characteristics suited to very dry habitats (p. 149), such as deserts. Xeromorphic plants often have thick, succulent (p. 99) leaves with thick cuticles (p. 95) to prevent the loss of water.

halophyte (*n*) a plant which lives in salty habitats (p. 149). Halophytes are adapted (p. 141) for the uptake of water from concentrated solutions (p. 12). **halophytic** (*adj*).

cryophyte (*n*) a plant growing in very cold conditions, e.g. on ice or snow. Cryophytes are usually algae (p. 119) or bryophytes (p. 122).

cryptophyte (*n*) a plant with buds (p. 110) or shoot apices (p. 90) perennating (p. 117) underground or under water during unfavourable seasons, e.g. winter.

geophyte (*n*) a cryptophyte (↑) with buds (p. 110) or shoot apices (p. 90) perennating (p. 117) underground. Geophytes have rhizomes (p. 60), bulbs (p. 60) or corms (p. 60), etc. **geophytic** (*adj*).

geophyte e.g. *Narcissus*

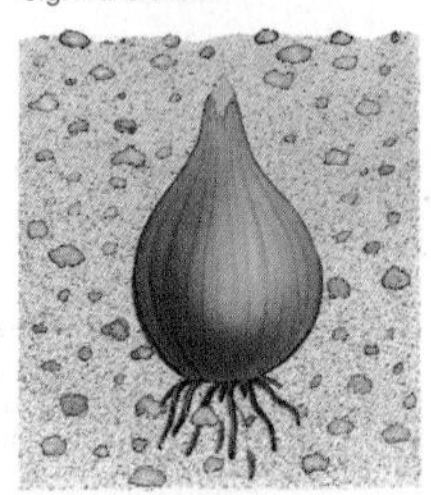

bulb perennating underground during winter

hydrophyte e.g. *Nuphar*

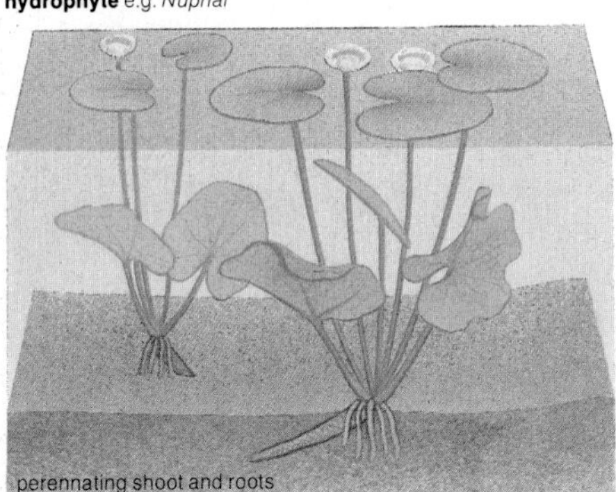

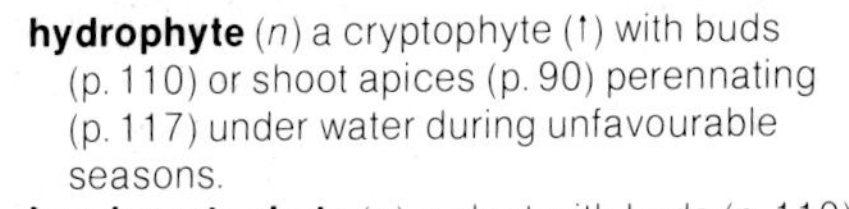

hydrophyte (*n*) a cryptophyte (↑) with buds (p. 110) or shoot apices (p. 90) perennating (p. 117) under water during unfavourable seasons.

hemicryptophyte (*n*) a plant with buds (p. 110) perennating (p. 117) at the surface of the soil.

chamaephyte (*n*) a woody or herbaceous (p. 136) plant less than 25 cm tall, with buds (p. 110) perennating (p. 117) above the ground.

phanerophyte (*n*) a woody plant with buds (p. 110) perennating (p. 117) more than 25 cm above the surface of the ground.

evolution (*n*) the changes that occur in organisms over many generations (p. 63) and long periods of time. Evolution occurs by the natural selection (p. 140) of mutations (p. 54). **evolutionary** (*adj*), **evolve** (*v*).

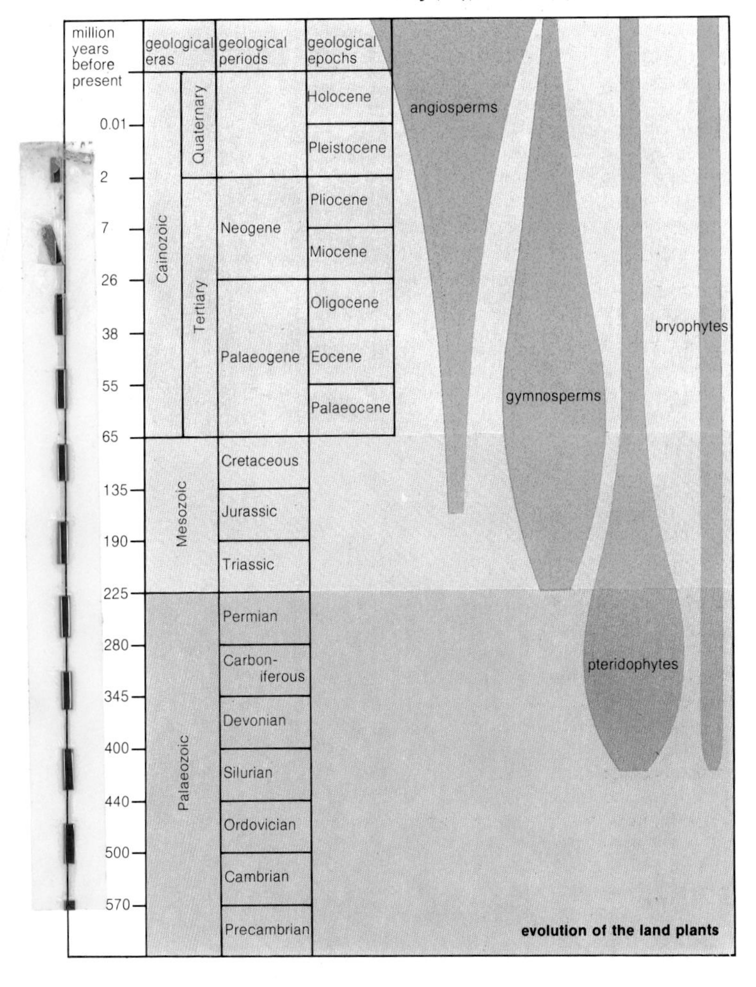

evolution of the land plants

Lamarck Jean-Baptiste Lamarck (1744-1829) proposed that evolution (p. 139) occurred as a result of the inheritance (p. 41) of changes which happen to an organism during its lifetime. This is known as the Theory of Inheritance of Acquired Characters.

Darwin Charles Darwin (1809-1882) proposed that evolution (p. 139) occurs by natural selection (↓) and the survival of the fittest (↓). His book on the subject, *On the Origin of Species*, was published in 1858. Darwin's theory of evolution is often known as Darwinism.

survival of the fittest the ability of some individuals (p. 135) in a population (p. 135) to live until after they have reproduced (p. 59), because they have characters which enable them to survive for longer than those individuals which die before reproduction. This is natural selection (↓).

natural selection Darwin's (↑) theory of evolution (p. 139), also called the survival of the fittest (↑). Natural selection is the selection by the environment (p. 149) of the individuals (p. 135) in a population (p. 135) which are fittest. Only these individuals live until reproductive (p. 59) age, and pass their genes (p. 41) to future generations (p. 63).

artificial selection the process by which man selects plant varieties (p. 135) or individuals (p. 135) with useful characters or traits (p. 41), e.g. wheat plants with large seeds, in order to breed (p. 59) them for his own purposes. Because of this, many cultivated plants appear to be very different from their wild ancestors.

neo-Darwinism (*n*) the theory of evolution (p. 139) developed in the 20th century, after Darwin's (↑) death. It includes Darwin's theory of natural selection (↑) and the more recent knowledge of genetics (p. 41) and inheritance (p. 41) through chromosomes (p. 46).

telome theory a theory of the evolution (p. 139) of the organs (p. 88) of plants, especially the branches and leaves, starting with very simple leafless primitive (↓) pteridophytes (p. 126) with green stems.

neoteny (*n*) the condition in which some of the characters of an embryonic (p. 85) or young organism are found in the mature organism of reproductive (p. 59) age. Neoteny may have been important in the evolution (p. 139) of flowering plants (p. 130).

phylogeny (*n*) the evolutionary (p. 139) history of an organism or taxonomic (p. 133) group of organisms. **phylogenetic** (*adj*).

primitive (*adj*) of organisms which have the characteristics of early stages in evolutionary (p. 139) history. This can also apply to characters and organs (p. 88).

extinct (*adj*) of species (p. 134) which no longer exist.

extant (*adj*) of species (p. 134) which exist at present.

adaptation (*n*) a character or set of characters of an organism which help the organism to survive and reproduce (p. 59) in a particular habitat (p. 149). **adapt** (*v*), **adaptive** (*adj*).

adaptive radiation a process of evolution (p. 139) from one primitive (↑) species (p. 134) to more than one advanced species, each adapted (↑) to a particular niche (p. 153) or habitat (p. 149).

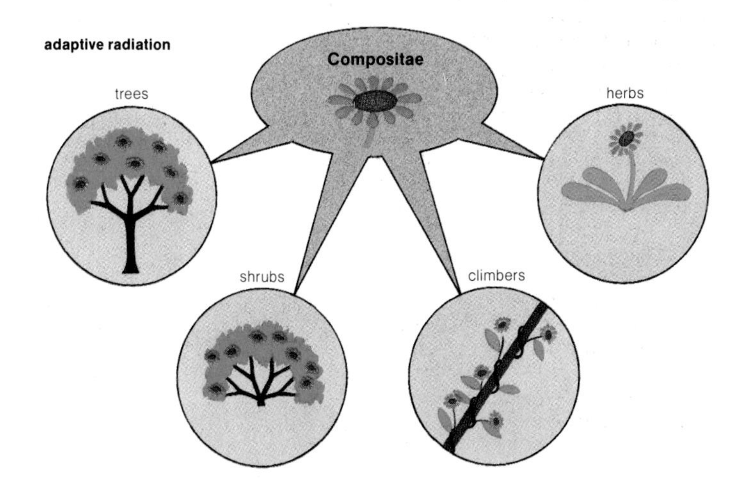

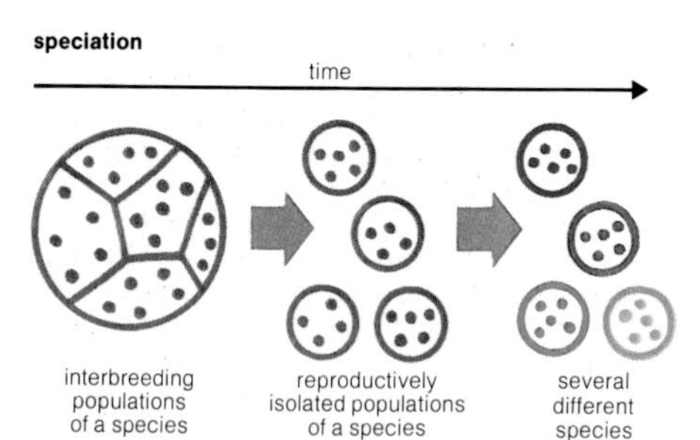

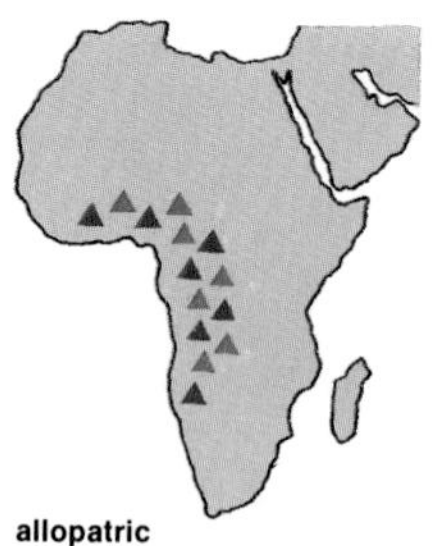

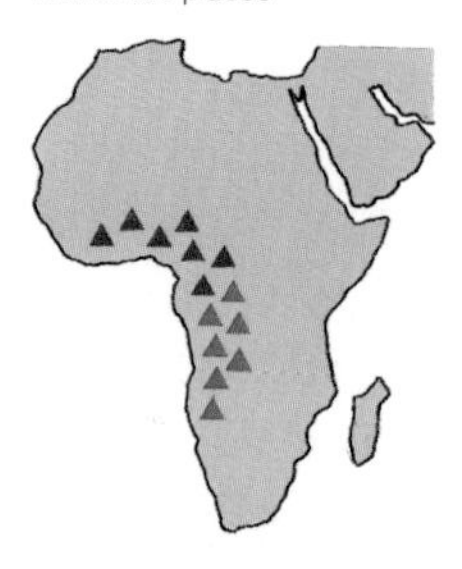

speciation (*n*) the evolutionary (p. 139) process in which new species (p. 134) are produced. **speciate** (*v*).

sympatric (*adj*) of two species (p. 134) which live in the same place, or of speciation (↑) in which two or more species evolve (p. 139) in the same place. **sympatry** (*n*).

allopatric (*adj*) of two or more species (p. 134) living in separate places, or of speciation (↑) in which different species evolve (p. 139) in different places. **allopatry** (*n*).

reproductive isolation a situation in which two individuals (p. 135) or populations (p. 135) cannot breed (p. 59) with each other.

fossil (*n*) the remains or marks left by dead organisms, converted to stone over geological time (↓). Fossils provide important clues to the history and evolution (p. 139) of living organisms. **fossilize** (*v*).

palaeobotany (*n*) the study of plants of the past, using fossils (↑).

palynology (*n*) the study of pollen (p. 74) remains and fossil (↑) pollen. This is an important way of studying the history of vegetation (p. 150). Pollen grains can last for thousands or millions of years, and it is possible to identify their genus (p. 134) or family (p. 134) by their shape and surface patterns.

pollen diagram a diagram showing the variation in the pollen (p. 74) content of sediments over long periods of time (often tens of thousands of years). Pollen diagrams are used to determine the history of vegetation (p. 150).

geological time the sequence of geological eras (↓) in the earth's history, measured in millions of years. Geological time began about 4½ thousand million years ago, when the earth was formed. *See also* page 139.

geological era a very long division of geological time (↑), lasting tens of millions of years, whose beginning and end are recognized by major changes in layers of rocks and fossils (↑) in the earth. Geological eras are divided into geological periods (↓) and epochs (↓).

geological period a major subdivision of a geological era (↑).

geological epoch a subdivision of a geological period (↑).

Palaeozoic (*n*) the geological era (↑) which ended about 225 million years ago, before the evolution (p. 139) of the angiosperms (p. 130). Pteridophytes (p. 126) were the dominant (p. 150) plants on land during this time.

Carboniferous (*n*) the geological period (↑) 345–280 million years ago, during the Palaeozoic (↑). The world's forests (p. 158) were dominated (p. 150) by tree-like pteridophytes (p. 126) which were fossilized (↑) to form coal.

Mesozoic (*n*) the geological era (↑) 225–65 million years ago, following the Palaeozoic (↑), with gymnosperms (p. 128) as the dominant (p. 150) plants on land. Angiosperms (p. 130) first appeared during the Mesozoic.

Cainozoic (*n*) the geological era (↑) which began 65 million years ago, after the Mesozoic (↑), and is still continuing. Angiosperms (p. 130) have been the dominant (p. 150) plants on land throughout this time. Also known as **Cenozoic**

Tertiary (*n*) one of two geological sub-eras (↑) into which the Cainozoic (↑) is divided. It began about 65 million years ago and ended about 2 million years ago.

Quaternary (*n*) the second geological sub-era (↑) of the Cainozoic (↑). It began with the Pleistocene (↓) and has lasted until the present.

Pleistocene (*n*) the geological epoch (↑) which began about 2 million years ago and ended 10 thousand years ago with the last Ice Age.

interaction (*n*) the process in which two or more organisms (p. 118), of the same or different species, act on each other, e.g. symbiosis (↓). **interact** (*v*).

symbiosis (*n*) the state of two different organisms living closely together for much or all of their lives, e.g. the fungi (p. 163) and algae (p. 119) in lichens (p. 147). **symbiotic** (*adj*).

symbiont (*n*) an organism living in a symbiotic (↑) relationship with another organism.

mutualism (*n*) the kind of symbiosis (↑) in which each partner in the relationship gains something from the other.

commensalism (*n*) the kind of symbiosis (↑) in which the partners live in close association with each other and neither partner gains an obvious advantage.

pathogen (*n*) an organism (p. 118) which causes disease or illness in another organism. Many viruses (p. 118), fungi (p. 163) and bacteria (p. 119) are pathogens. **pathogenic** (*adj*).

infection (*n*) the entry or presence of a parasite (↓), pathogen (↑) or symbiont (↑) in the tissue (p. 88) of a host (↓). **infect** (*v*).

phytopathology (*n*) the study of the diseases of plants.

parasite (*n*) an organism which takes all its nutrients (p. 111) from the tissues (p. 88) of another organism, usually with harmful effects. Many fungi (p. 163) and bacteria (p. 119) are parasites, and so are a few flowering plants (p. 130), e.g. dodder, broomrape. **parasitic** (*adj*).

hemiparasite (*n*) a green plant whose roots grow into the tissues (p. 88) of another plant. Hemiparasites photosynthesize (p. 32), but take some of their nutrients (p. 111) and water from the other plant, e.g. mistletoe. Also known as semi-parasites.

vector[2] (*n*) an animal which carries parasites (↑) or pathogens (↑) from one organism to another.

host (*n*) the general name for an organism with a parasite (↑) in it, a plant with an epiphyte (p. 137) on it, or the larger partner in a symbiotic (↑) relationship.

pathogens

e.g. basidiomycete fungus causing rust on wheat leaves

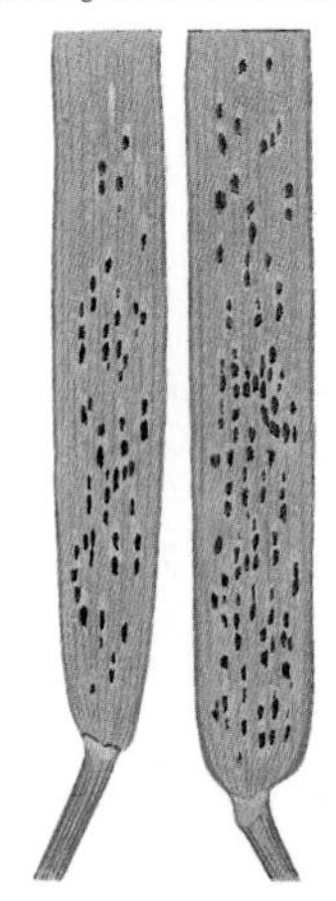

e.g. bacteria causing galls on apple tree stems

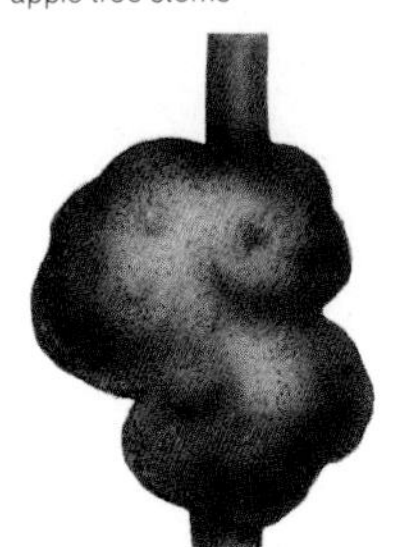

hemiparasite

mycorrhizae

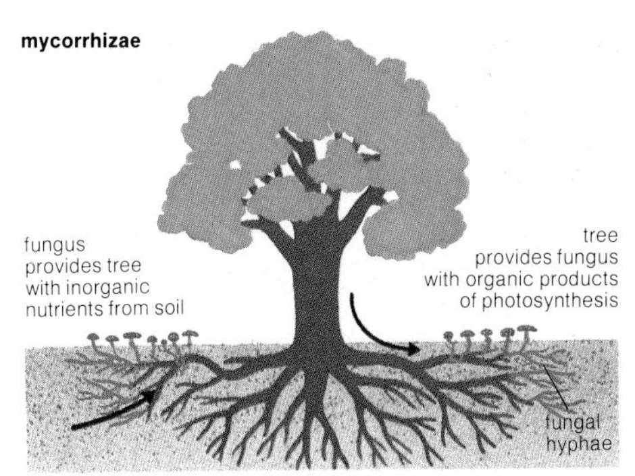

mycorrhiza (*n*) a symbiotic (↑) association between a fungus (p. 163) and the underground parts of a plant. Plants with mycorrhizae (*pl.*) often grow faster or larger than plants without, because the mycorrhizae provide extra nutrients (p. 111). In most cases, the plant provides the mycorrhiza with products of photosynthesis (p. 32). Mycorrhizae are common in most plant families (p. 134).

ectotrophic mycorrhiza

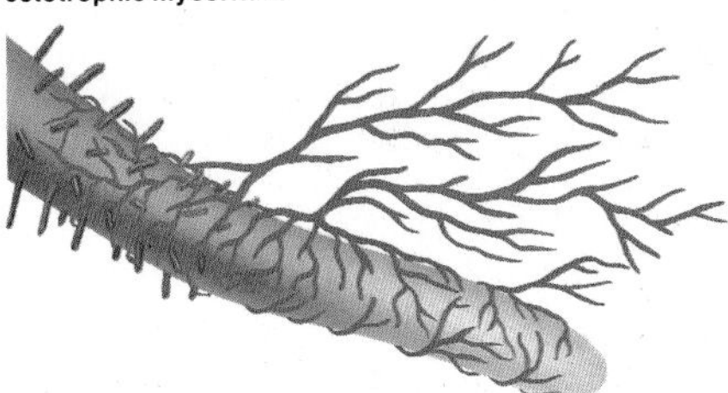

endotrophic mycorrhiza

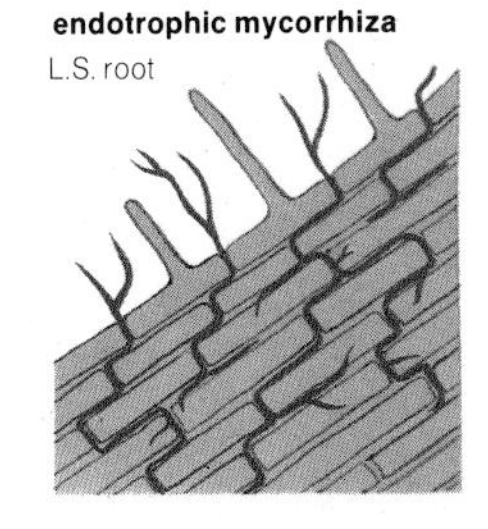

ectotrophic (*adj*) of mycorrhizae (↑) whose hyphae (p. 163) do not grow into the cells of the host (↑). Ectotrophic mycorrhizae form a sheath (p. 100) around the root of the host, and the mycelium (p. 163) also grows in the intercellular spaces (p. 95) of the root tissues (p. 88) *See also* **endotrophic** (↓).

endotrophic (*adj*) of mycorrhizae (↑) which do not form a sheath (p. 100) around the root of the host (↑). Endotrophic mycorrhizae usually grow into the cells of the host.

haustorium

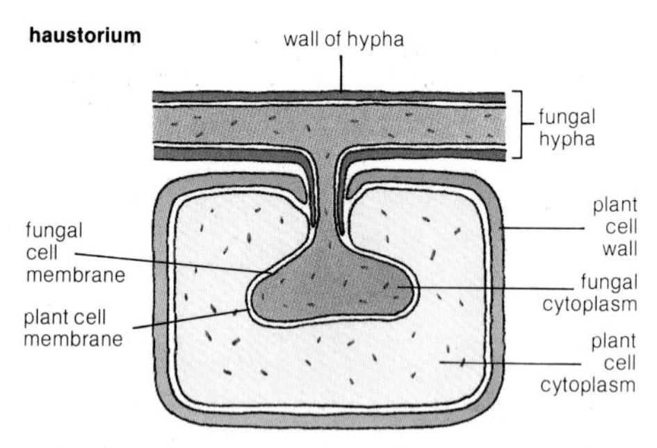

haustorium (*n*) the part of the hypha (p. 163) of a parasitic (p. 144) fungus (p. 163) which grows into the cell of the host (p. 144). **haustoria** (*pl.*).

nodule (*n*) a swelling on the root of a member of the family (p. 134) Leguminosae (p. 131), caused by symbiotic (p. 144) *Rhizobium* bacteria (p. 119). These bacteria are involved in nitrogen fixation (↓).

root nodules

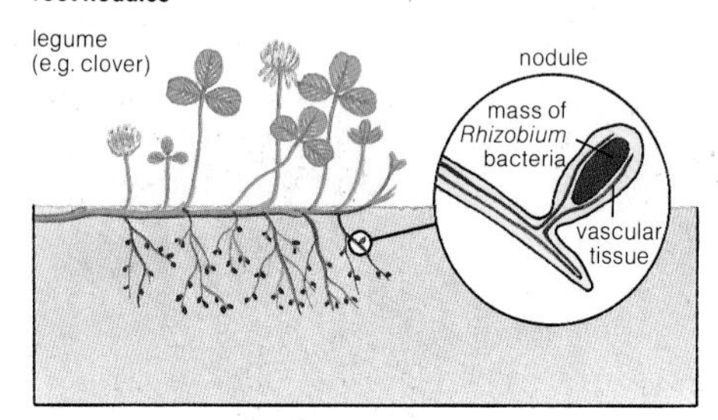

nitrogen fixation the process in which nitrogen in the air (N_2) is reduced (p. 11) by organisms to ammonia (p. 13). Only prokaryotic (p. 16) organisms such as blue-green algae (p. 120) and bacteria (p. 119) can carry this out. Some nitrogen-fixing organisms are found in symbiotic (p. 144) relationships, e.g. blue-green algae in lichens (↓) or *Rhizobium* bacteria in root nodules (↑).

gut flora the microorganisms (p. 118) found in the gut of animals. The gut flora assists the animal in breaking down its food.

lichen (*n*) a symbiosis (p. 144) between a green or blue-green alga (p. 120) and a fungus (p. 163). Lichens are usually small plants, with a range of colour, which grow on rocks or as epiphytes (p. 137).

phycobiont (*n*) the algal (p. 119) symbiont (p. 144) in a lichen (↑).

mycobiont (*n*) the fungal (p. 163) partner in a lichen (↑).

fruticose (*adj*) of lichens (↑) whose habit (p. 163) is shrub-like (p. 136).

foliose (*adj*) of lichens (↑) with a leafy thallus (p. 122). Foliose lichens have distinct upper and lower surfaces.

crustose (*adj*) of lichens (↑) whose thallus (p. 122) is closely pressed to the substrate (p. 154) or actually growing within it.

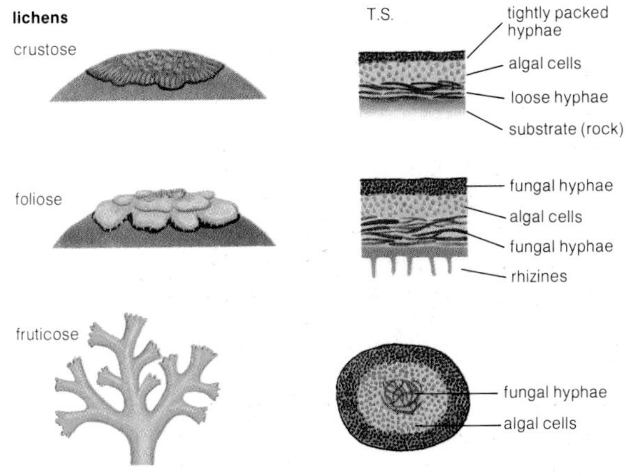

rhizine (*n*) a root-like bunch of hyphae (p. 163) growing from the bottom of the thallus (p. 122) of a lichen (↑).

apothecium (*n*) a cup-shaped structure which bears spores (p. 66), found in some lichens (↑).

perithecium (*n*) a flask-shaped hollow structure which bears spores (p. 66), in some lichens (↑). The perithecium opens through a pore (p. 19) in the surface of the thallus (p. 122).

obligate (*adj*) of organisms that can only live in one way. For instance, the fungi (p. 163) and algae (p. 119) in lichens (p. 147) are obligate symbionts (p. 144), being unable to live without each other, in most cases.

facultative (*adj*) of organisms which can live under several kinds of conditions. For instance, a facultative epiphyte (p. 137) is a plant which can grow either on the ground or on other plants.

toxin (*n*) a substance which is poisonous. Plants produce toxins, e.g. alkaloids (↓), to protect themselves from attack by herbivores (p. 153). **toxic** (*adj*).

phytoalexin (*n*) a substance produced in some plants, which prevents attack by pathogenic (p. 144) or parasitic (p. 144) fungi (p. 163).

antibiotic (*n*) a substance that is harmful to bacteria (p. 119). Antibiotics are produced by many fungi (p. 163), e.g. penicillin is produced by several species of the fungus *Penicillium*.

tannins (*n.pl.*) a group of substances common in the outer tissues (p. 88) of many plants, which are bitter to the taste and are a defence against herbivores (p. 153). Tannins are used in the tanning of leather.

alkaloid (*n*) an organic (p. 11), nitrogen-containing compound, produced by many plants. Alkaloids are mostly toxic (↑) and often defend plants against attack by herbivores (p. 153).

allelopathy (*n*) discouragement by one plant of the growth of other plants around it, e.g. by toxins (↑) contained in the fallen leaves. **allelopathic** (*adj*).

insectivorous (*adj*) of organisms which eat insects. Some plant species (p. 134) trap insects with sticky hairs (e.g. sundew), in bucket-shaped leaves (e.g. pitcher plants) or between hinged leaf-lobes (e.g. Venus fly-trap). Insectivorous plants obtain nutrients (p. 111) by secreting (p. 112) enzymes (p. 15) which break down the tissues (p. 88) and cells of the trapped insect. The insectivorous habit (p. 136) is an adaptation (p. 141) to habitats (↓) that are poor in nitrates (p. 13).

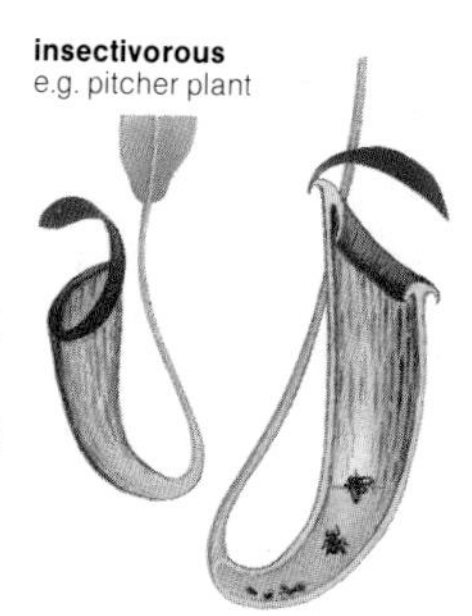
insectivorous
e.g. pitcher plant

ecology (*n*) the study of organisms in relation to their environment (↓). **ecological** (*adj*), **ecologist** (*n*).
autecology (*n*) the ecology (↑) of a single species (p. 134) in a habitat (↓).
synecology (*n*) the ecology (↑) of all the organisms found in a habitat (↓) or an ecosystem (↓).

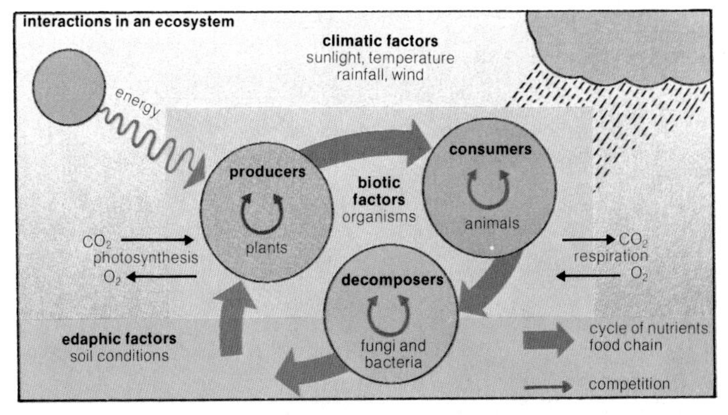

ecosystem (*n*) an ecological (↑) system in which organisms interact (p. 144) with each other and with their non-living environment (↓) and in which there is a more or less closed cycle of nutrients (p. 111).
biosphere (*n*) the parts of the earth in which organisms live, including the oceans, the land, the soil, and the atmosphere.
environment (*n*) the living and non-living surroundings of an organism, and the events which take place in those surroundings. **environmental** (*adj*).
habitat (*n*) the place or kind of place in which an organism, community (↓) or association (p. 150) is found, e.g. the habitat of an epiphyte (p. 137) is in the branches of trees, the habitat of algae (p. 119) is water.
community (*n*) the group of species (p. 134) of plants, animals, or both, living in the same habitat (↑) and interacting (p. 144) with each other.

association (*n*) a group of species (p. 134) that are usually found together, and require the same habitat (p. 149).
phytosociology (*n*) the study of the associations (↑) of plants.
dominant[2] (*adj*) of the most common or the largest species (p. 134) in a community (p. 149). **dominant** (*n*), **dominate** (*v*).
vegetation (*n*) the general term for all the plants in an ecosystem (p. 149).
primary vegetation vegetation (↑) which has not been disturbed or changed by man.
secondary vegetation vegetation (↑) growing in a place which has been disturbed by man, e.g. roadsides, old farmland, etc.

examples of regeneration

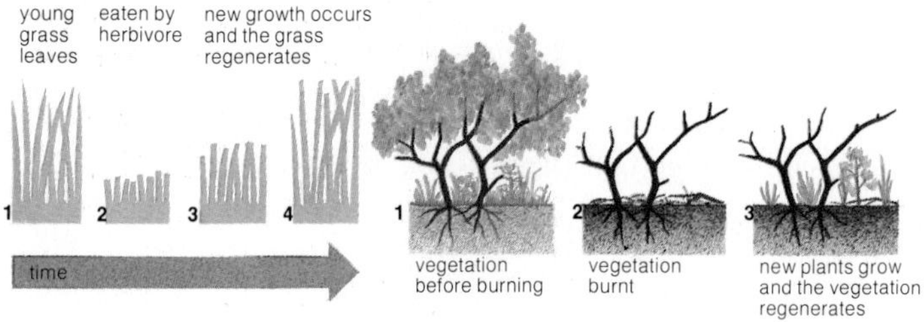

regeneration[2] (*n*) the growth of new vegetation (↑) in a place where the old vegetation has been removed or damaged.
ecotone (*n*) the border between two habitats (p. 149) or types of vegetation (↑).
producer (*n*) an autotrophic (p. 32) organism in an ecosystem (p. 149), which produces organic (p. 11) matter using chemical energy or energy from light. Plants are the main producers in the biosphere (p. 149).
primary production the total amount of organic (p. 11) matter produced by the autotrophic (p. 32) organisms in an ecosystem (p. 149), using energy from sunlight.
primary productivity the amount of matter that can be synthesized (p. 13) by all the autotrophic (p. 32) organisms in a given area in a given time.

ecotone

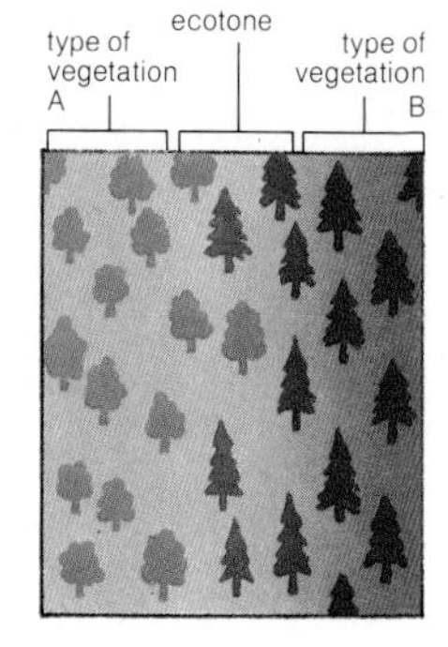

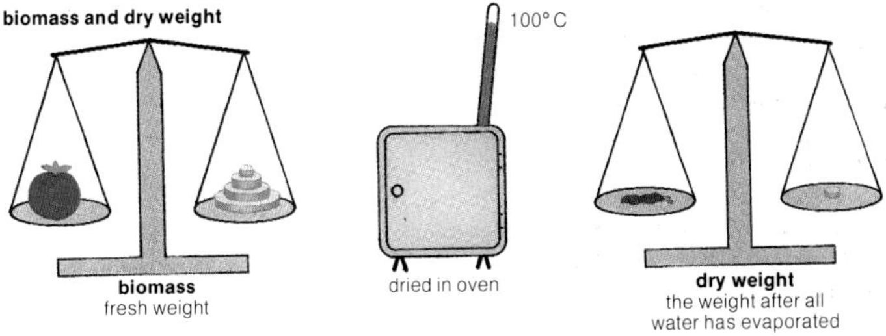

biomass (*n*) the weight of a living organism, or of all the organisms in an ecosystem (p. 149) or in a habitat (p. 149).

dry weight the weight of an organism, part of an organism, or the organisms in a habitat (p. 149) or in an ecosystem (p. 149), after drying. Because a large part of the biomass (↑) of most organisms is water, dry weight is usually small compared to biomass.

colonization (*n*) the arrival and germination (p. 87) of a seed on a substrate (p. 154), or the spread of plants to places where they have not grown before. Successful colonization depends on growth to reproductive (p. 59) age.

pioneer (*n*) a plant species (p. 134) that occurs in the early stages of succession (↓).

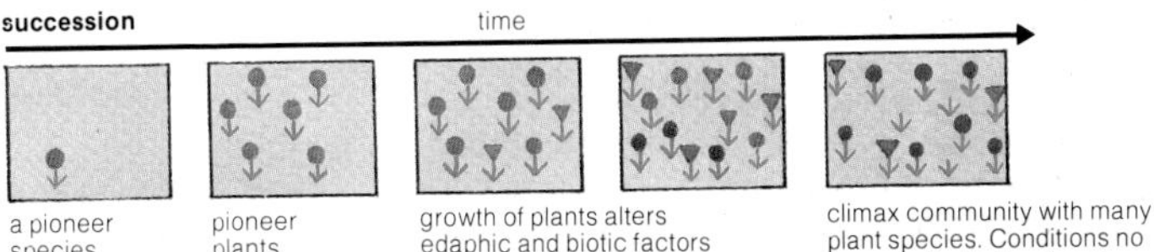

succession (*n*) the process of development of vegetation (↑), involving changes of species (p. 134) and communities with time. Succession occurs because the growth of plants alters the biotic factors (p. 152) and edaphic factors (p. 154) of a habitat (p. 149), making possible the colonization (↑) of other species. **successional** (*adj*).

open community a plant community (p. 149) in which the niches (↓) are unstable or 'empty', allowing entry into the community of new species (p. 134) from outside.

closed community a plant community (p. 149) in which the niches (↓) are stable and 'full', not allowing the entry of extra species (p. 134).

sere (*n*) a succession (p. 151) in a particular habitat (p. 149), e.g. a hydrosere is a succession in a shallow aquatic (p. 161) habitat, beginning with aquatic plants and ending with swamp forest (p. 158).

climax (*n*) the last stage in a succession (p. 151), after which there are no further great changes in the structure or the species (p. 134) in a habitat (p. 149).

biotic factors the effects of living organisms on an ecosystem (p. 149) and on each other, e.g. herbivores (↓) eating plants, or trees casting shade.

competition

intraspecific
between individuals of the same species in a habitat

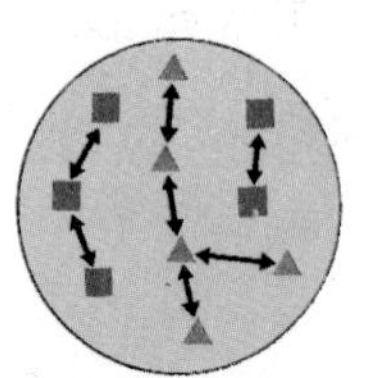

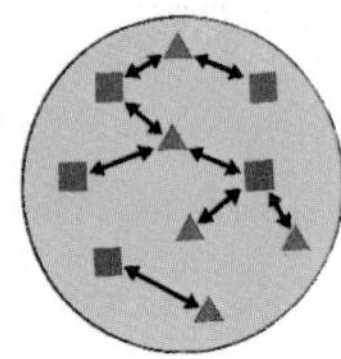

interspecific
between individuals of different species in a habitat

competition (*n*) an interaction (p. 144) between two or more organisms in the same habitat (p. 149), when both, partly or wholly, share the same needs. Competition can be intraspecific (↓) or interspecific (↓). **compete** (*v*), **competitor** (*n*).

interspecific (*adj*) of competition (↑) between two species (p. 134).

intraspecific (*adj*) of competition (↑) between individuals (p. 135) of the same species (p. 134).

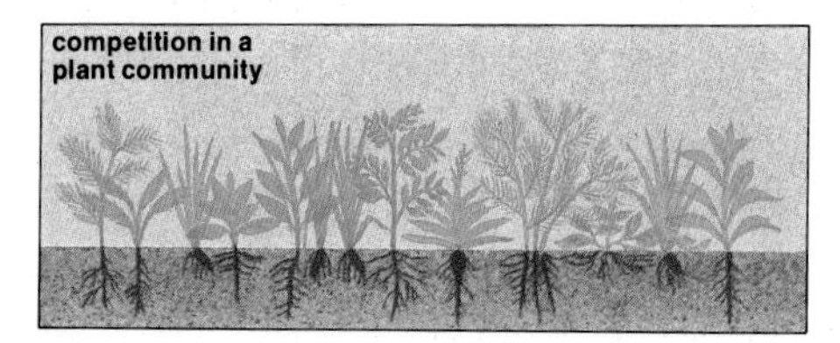

leaves compete for light

roots compete for nutrients

a food chain

producer
plant

consumer
herbivore

decomposer
bacteria

niche (*n*) the position and activities of an organism in its habitat (p. 149). Each species (p. 134) has its own niche, and competition (↑) occurs when the niches overlap.

herbivore (*n*) an animal that eats plants. **herbivory** (*n*), **herbivorous** (*adj*).

food chain the flow of energy and nutrients (p. 111) from one group of organisms to another in an ecosystem (p. 149), e.g. from producers (p. 150) to consumers (p. 154) to decomposers (p. 157).

a food web

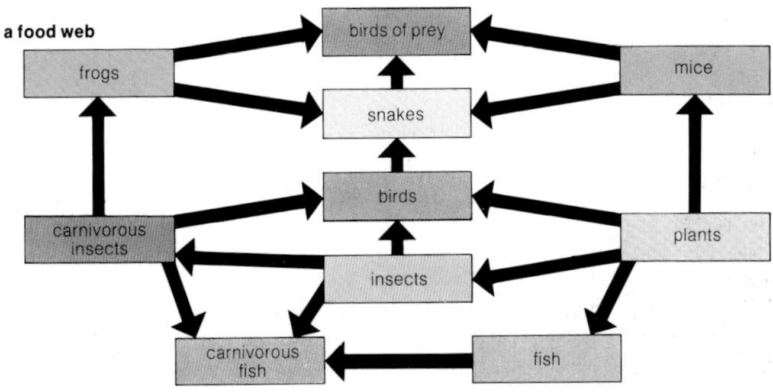

pyramid of available energy at the trophic levels of a food web

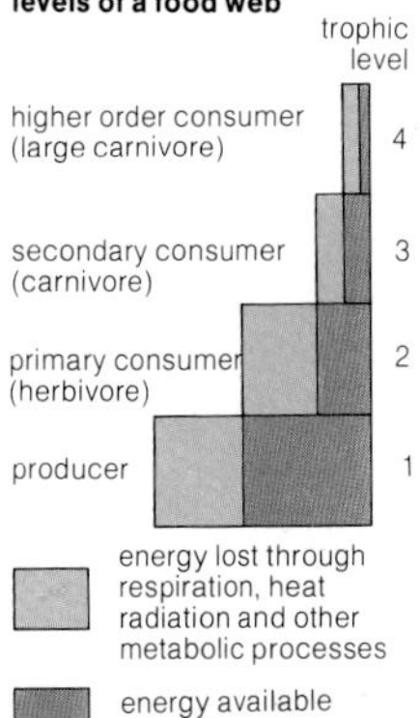

food web a set of interacting (p. 144) food chains (↑), e.g. an animal may feed on several plant species (p. 134), and the animal may be fed on by several animal species which in turn may be fed on by other animal species.

pyramid of numbers the number of organisms at each trophic level (↓) in a food web (↑) or food chain (↑). At each level, energy is lost through respiration (p. 22) and other metabolic (p. 14) processes, and there is less energy available for the next trophic level. For this reason, the number and biomass (p. 151) of consumers (p. 154) in an ecosystem (p. 149) are less than the number and biomass of producers (p. 150).

trophic level the position of an organism in a food chain (↑). The main trophic levels are those of producers (p. 150), consumers (p. 154) and decomposers (p. 157).

consumer (*n*) a heterotrophic (p. 32) organism which eats other organisms, e.g. a herbivore (p. 153).
carbon cycle the pathway of the element (p. 8) carbon through ecosystems (p. 149). Carbon dioxide is fixed from the atmosphere by plants during photosynthesis (p. 32) and used for the synthesis of organic (p. 11) compounds. These are passed through the food web (p. 153) and metabolized (p. 14) by animals and decomposers (p. 157), and carbon dioxide is released back to the atmosphere by respiration (p. 22).
nitrogen cycle the cycle of the element (p. 8) nitrogen through ecosystems (p. 149). Organisms need nitrogen in order to synthesize (p. 13) amino acids (p. 56), proteins (p. 56) and other nitrogen-containing organic (p. 11) compounds. Nitrogen is taken up from the soil by plants in the form of nitrate (p. 13), converted to plant protein, and may then be passed to animals as protein. It is returned to the soil due to the death and decay (p. 157) of plants and animals, and by animal excretion (p. 112). Decomposers (p. 157) in the soil break down the organic nitrogen-containing compounds to inorganic (p. 11) compounds, e.g. nitrate and ammonia (p. 13), thus completing the cycle. Atmospheric nitrogen (N_2) enters the cycle as a result of nitrogen-fixing (p. 146) organisms and by oxidation (p. 11) in lightning in thunderstorms.
nitrifying bacteria bacteria (p. 119) in the soil which oxidize (p. 11) ammonia (NH_3) (p. 13) to nitrate (NO_3^-) (p. 13). This is one of the important stages in the nitrogen cycle (↑), which makes nitrate available for plants.
denitrifying bacteria bacteria (p. 119) in the soil which reduce (p. 11) nitrate (NO_3^-) (p. 13) to nitrite (NO_2^-) (p. 13) and molecular nitrogen (N_2).
edaphic factors the effects of the soil in an ecosystem (p. 149). Different soils have different structural and chemical characteristics, and different plant species (p. 134) are adapted (p. 141) for growth on particular types of soil.
substrate[2] (*n*) general term for the soil or the surface on which an organism is growing.

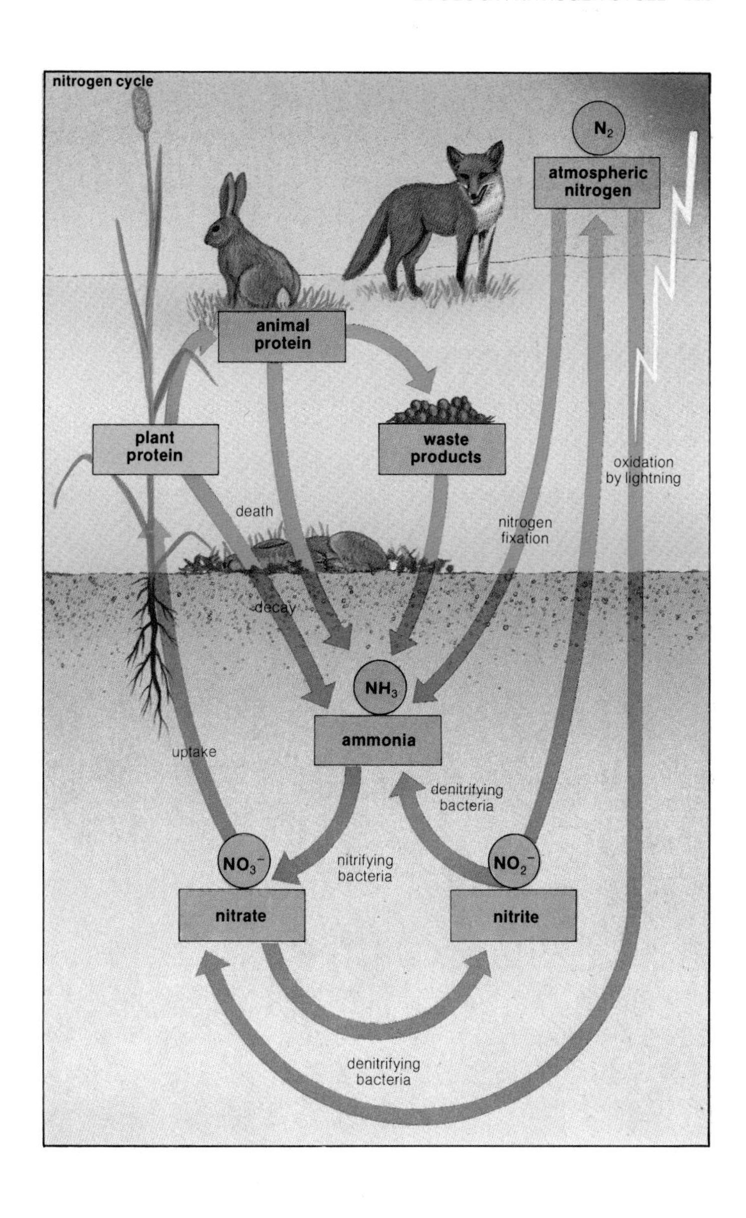
nitrogen cycle
N_2
atmospheric nitrogen
animal protein
plant protein
waste products
death
decay
nitrogen fixation
oxidation by lightning
NH_3
ammonia
uptake
denitrifying bacteria
NO_3^-
nitrate
nitrifying bacteria
NO_2^-
nitrite
denitrifying bacteria

soil profile the sequence of different layers of material in the soil. The layers, or horizons (↓), differ in chemical composition and thickness. The top layers are usually organic (p. 11), derived from the litter (↓), and the layers underneath are inorganic (p. 11), derived from the rock beneath them. Soils in different places have their own characteristic soil profiles, depending on the climate and the type of rock on which they occur.

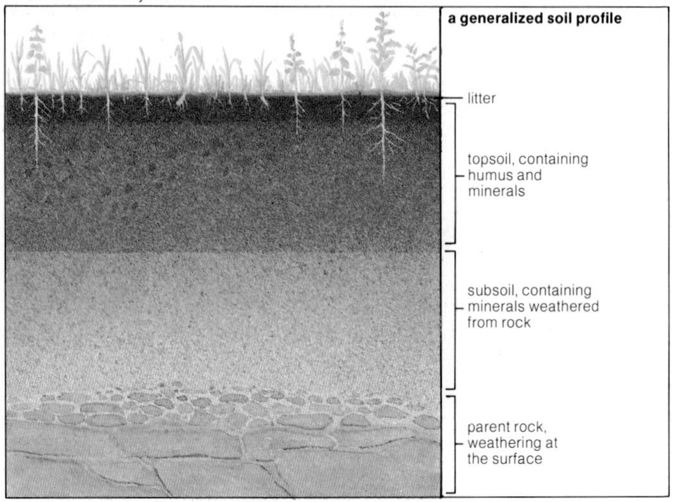

topsoil (*n*) the general term for the upper, organic (p. 11) horizons (↓) in a soil profile (↑).
subsoil (*n*) the general term for the lower, inorganic (p. 11) horizons (↓) in a soil profile (↑).
horizon (*n*) a layer in a soil profile (↑). Different soil profiles can be compared by examining the structure, thickness and chemical composition of their horizons.
humus (*n*) the layer of organic (p. 11) matter at the top of a soil profile (↑). Humus is the habitat (p. 149) of most decomposers (↓).
litter (*n*) dead plant and animal material on the surface of the ground, above the humus (↑) layer.

peat (*n*) a kind of litter (↑) layer found in certain very wet or waterlogged habitats (p. 149), such as bogs, which decomposes (↓) very slowly, often under very acid conditions. Layers of peat may be several metres thick.

mor (*n*) a very acid humus (↑) which hardly mixes with the inorganic (p. 11) soil underneath it.

mull (*n*) humus (↑) which is well-mixed with the inorganic (p. 11) soil.

calcareous (*adj*) of substrates (p. 154) which contain calcium carbonate, $CaCO_3$, e.g. soils on limestone or chalk.

calcicole (*n*) a plant which grows only or mainly on calcareous (↑) soil. **calcicolous** (*adj*).

calcifuge (*n*) a plant which grows only or mainly on non-calcareous (↑) soil. **calcifugous** (*adj*).

decomposer (*n*) an organism which breaks down organic (p. 11) matter, releasing carbon dioxide and inorganic (p. 11) compounds, e.g. nitrates (p. 13), phosphates (p. 13), ammonia (p. 13). The main decomposers are bacteria (p. 119) and fungi (p. 163). **decomposition** (*n*), **decompose** (*v*).

decay (*n*) the process of rotting and decomposition (↑) which takes place after the death of an organism. Decay involves the breakdown of organic (p. 11) compounds in the organism, by saprophytic (p. 137) bacteria (p. 119) and fungi (p. 163). It is an important part of the cycle of nutrients (p. 111) and energy in ecosystems (p. 149). **decay** (*v*).

rhizosphere (*n*) the general name given by some ecologists to the parts of the biosphere (p. 149) in which roots grow.

forest (*n*) a habitat (p. 149) or type of vegetation (p. 150) in which large trees are dominant (p. 150), forming a dense canopy (↓).

canopy (*n*) the top layer of a forest (↑), consisting of the crowns (p. 92) of trees.

understorey (*n*) the part of a forest (↑) or woodland (↓) underneath the canopy (↑), consisting of shrubs (p. 136), saplings (p. 136) and herbs (p. 136).

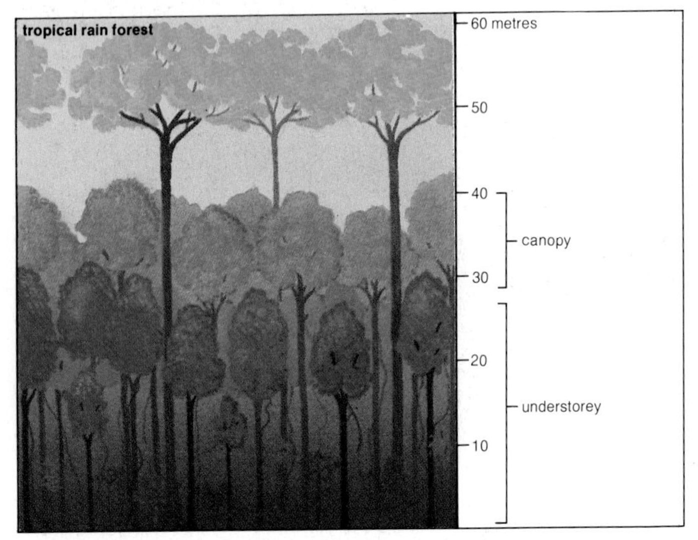

rain forest a wet forest (↑), where there is heavy rain during most months of the year. Most rain forests are tropical (p. 162), although some are found in mild rainy temperate (p. 162) regions. Tropical rain forests usually have very tall trees and a very large number of plant species (p. 134).

jungle (*n*) thick, secondary vegetation (p. 150) in wet tropical (p. 162) regions.

montane forest a forest (↑) on a mountain. Montane forests have smaller trees than lowland forests, and the trees get smaller towards the tree line (↓).

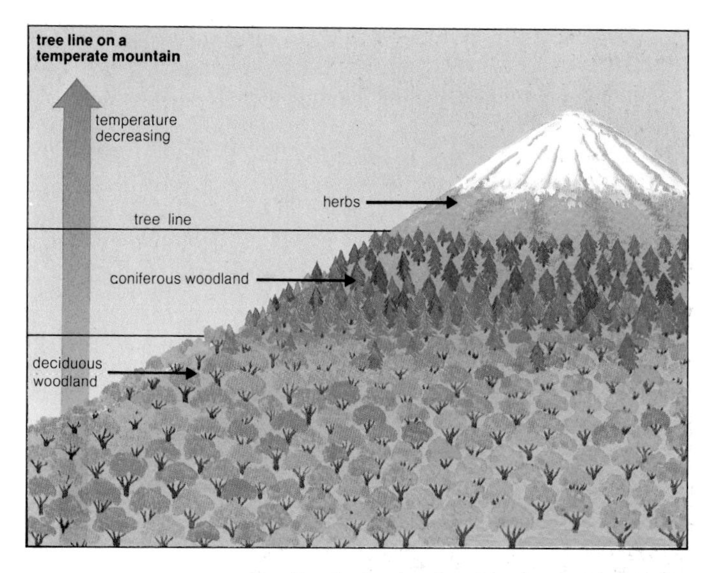

tree line the level on the side of a mountain above which trees do not grow. Vegetation (p. 150) at higher levels is herbaceous (p. 136) or absent altogether.

woodland (*n*) vegetation (p. 150) in which trees are dominant (p. 150). The trees in woodland are smaller and more spaced apart than those in forest (↑).

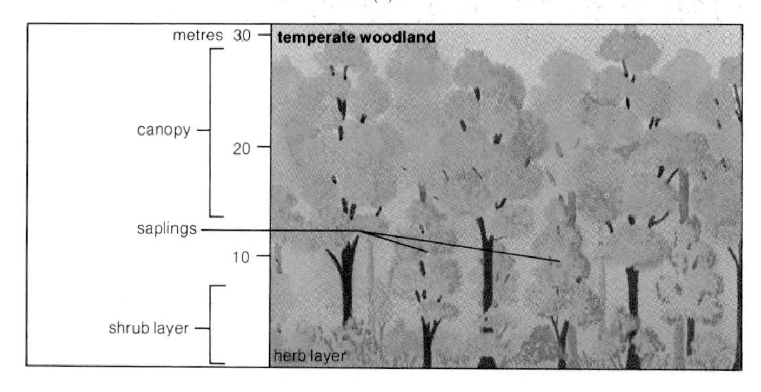

scrub landscape dominated by shrubs

scrub (*n*) vegetation (p. 150) in which shrubs (p. 136) and small trees are dominant (p. 150).
grassland (*n*) vegetation (p. 150) in which grasses (p. 130) are dominant (p. 150), e.g. prairies (↓), savanna (↓).

grassland landscape dominated by grasses

prairie (*n*) a North American grassland (↑).
savanna (*n*) a tropical (p. 162) grassland (↑).
steppes (*n.pl.*) the grasslands (↑) of temperate (p. 162) Asia.
sward (*n*) an area of vegetation (p. 150) consisting mainly of grasses (p. 130).

aquatic habitats

freshwater

sea water

brackish water

salt marsh

littoral zone

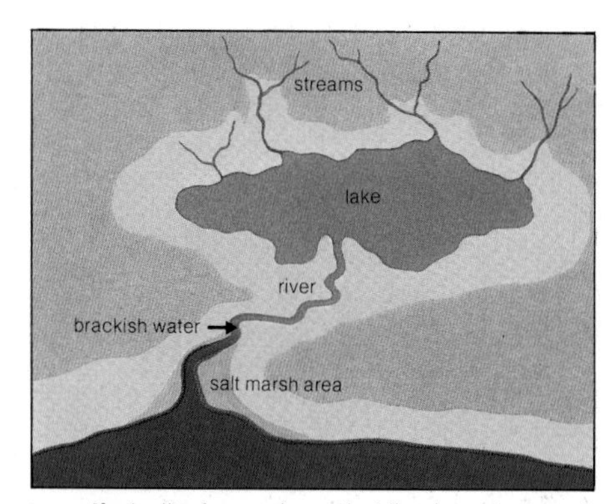

aquatic (*adj*) of organisms that live in water, or of underwater habitats (p. 149).

freshwater (*adj*) of aquatic (↑) habitats (p. 149) in which the concentration (p. 12) of dissolved (p. 12) ions (p. 9) is very low, e.g. rivers, streams, lakes.

eutrophic (*adj*) of habitats (p. 149) rich in nutrients (p. 111).

oligotrophic (*adj*) of habitats (p. 149) poor in nutrients (p. 111).

eutrophication (*n*) a process which can occur in rivers and shallow lakes when the addition of extra nutrients (p. 111), e.g. from fertilizers, causes heavy growth of algae (p. 119). When the algae die, their decay (p. 157) by bacteria (p. 119) reduces the concentration (p. 12) of oxygen in the water, so that aerobic (p. 22) organisms may not survive.

brackish water water in which the concentration (p. 12) of dissolved (p. 12) ions (p. 9) is more than in freshwater (↑) habitats (p. 149) but less than in seawater.

salt marsh a coastal habitat (p. 149) with a wet substrate (p. 154) containing a high concentration (p. 12) of dissolved (p. 12) salts, due to flooding by very high tides. The vegetation (p. 150) of salt marshes is mostly herbaceous (p. 136).

littoral (*adj*) of habitats (p. 149) between the high and low tide marks on the sea shore.

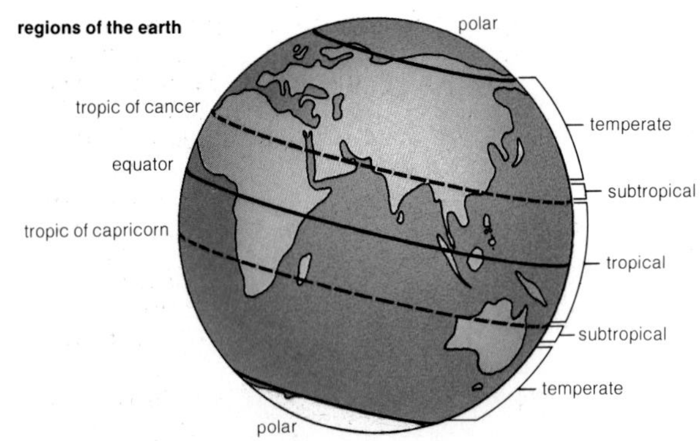

climatic factors the effects of temperature, sunlight, rainfall, wind, etc. on ecosystems (p. 149).

tropical (*adj*) of the regions of the world where the average monthly temperature changes little during the year, and the length of the day changes only slightly at different times of year.

subtropical (*adj*) of the regions of the world between the tropical (↑) and temperate (↓) regions.

temperate (*adj*) of the regions of the world which have warm summers with long days and cool winters with long nights.

polar (*adj*) of the very cold regions of the world close to the north and south poles, where the sun does not rise in midwinter and does not set in midsummer. Hardly any plants survive in polar regions.

phenology (*n*) the study of organisms and their activities in relation to the seasons of the year.

microclimate (*n*) the climate of a small or limited space, e.g. the surface of the soil, or under the canopy (p. 158) of a forest (p. 158).

quadrat (*n*) a square area marked out by an ecologist (p. 149) for counting and sampling organisms in a habitat (p. 149).

transect (*n*) a long rectangular area or a set of quadrats in a line marked out by an ecologist (p. 149) for counting and sampling organisms in one or several habitats (p. 149).

fungi (*n.pl.*) the large group of organisms, sometimes regarded as a separate kingdom (p. 134), which are set apart from other plants because they are heterotrophic (p. 32), lack chlorophyll (p. 36), and often have chitin (↓) in their cell walls (p. 17). Most fungi consist of thread-like hyphae (↓), which together form the mycelium (↓), although some, like yeast (p. 164), are unicellular (p. 119). Fungi reproduce (p. 59) by spores (p. 66). They are important as decomposers (p. 157) in ecosystems (p. 149), and many are parasitic (p. 144). **fungus** (*sing.*), **fungal** (*adj*).

mycology (*n*) the study of fungi (↑). **mycologist** (*n*).

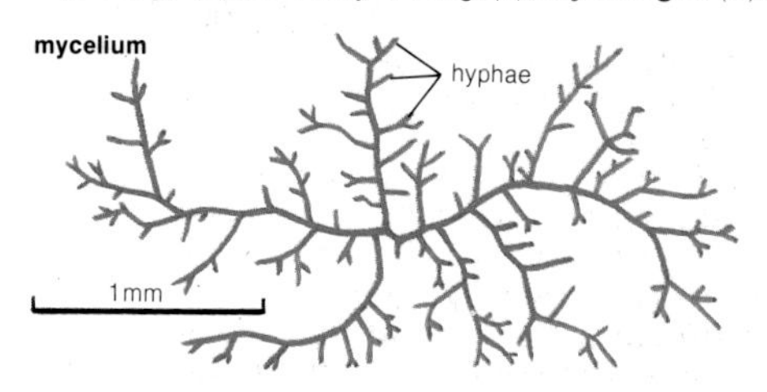

mycelium (*n*) the vegetative (p. 60) part of a fungus (↑), consisting of many hyphae (↓). **mycelia** (*pl.*).

hypha (*n*) a thread-like multinucleate (↓) tube with a cell wall (p. 17), the organ (p. 88) of vegetative (p. 60) growth (p. 109) in most fungi (↑). Hyphae increase in length by growth at their tips and give rise to new hyphae by side-branching. **hyphae** (*pl.*).

chitin (*n*) a polysaccharide (p. 30) containing nitrogen. This is the main substance in most fungal (↑) cell walls (p. 17). Chitin also occurs in insects.

multinucleate (*adj*) of cells with many nuclei (p. 19), as in the hyphae (↑) of fungi (↑).

septum (*n*) a wall across a hypha (↑). The number of nuclei (p. 19) between each septum varies from one or two, as in Basidiomycetes (p. 165), to many, as in other groups. **septa** (*pl.*).

aseptate (*adj*) of hyphae (↑) without septa (↑), e.g. in the Phycomycetes (p. 164).

Phycomycetes (*n*) a group of simple, aseptate (p. 163) fungi (p. 163), living mainly in damp conditions; the hyphae (p. 163) of Phycomycetes are not usually organized into mycelia (p. 163).

mildew (*n*) a disease of plants caused by a fungus (p. 163) growing on the surface. There are two common types of mildew, downy and powdery, which are produced by different kinds of fungi.

mould (*n*) the general name for a fungal (p. 163) growth on a surface.

asci and ascospores of Ascomycetes

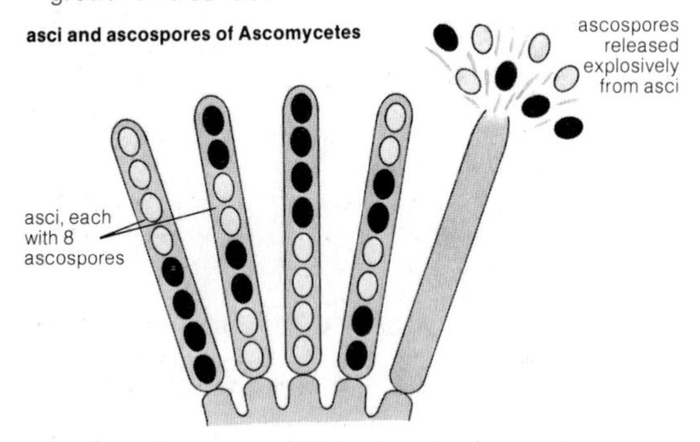

Ascomycetes (*n*) a large group of fungi (p. 163), recognized by their production of asci (↓) and ascospores (↓), e.g. yeast (↓).

ascospore (*n*) the haploid (p. 50) spore (p. 66) of an ascomycete (↑), formed by meiosis (p. 49) immediately after nuclear fusion (p. 61). Ascospores are contained within asci (↓), from which they are violently ejected when ripe (p. 83).

ascus (*n*) the reproductive (p. 59) organ (p. 88) of an ascomycete (↑), usually containing 8 ascospores (↑). Asci (*pl.*) are usually long and thin, with the ascospores arranged in a row.

yeast (*n*) kind of ascomycete (↑) fungus (p. 163). Yeasts, e.g. *Saccharomyces*, are unicellular (p. 119) and do not produce hyphae (p. 163) or mycelia (p. 163). Yeast cells can reproduce by budding. Yeasts are used by man for baking and brewing.

a budding yeast cell

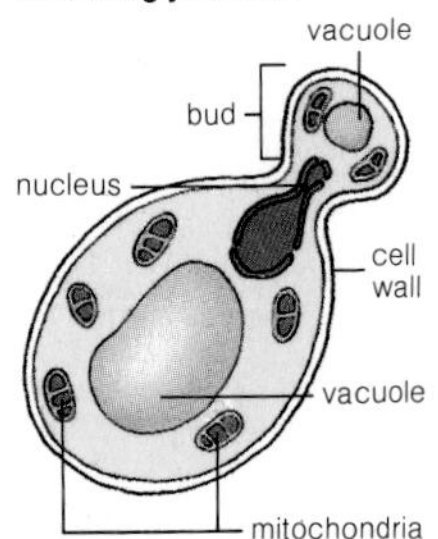

Deuteromycetes (*n*) a group of fungi (p. 163) that only reproduce (p. 59) asexually (p. 59). They are common and some, e.g. *Penicillium*, are useful to man. Also known as Fungi Imperfecti (↓).
Fungi Imperfecti = Deuteromycetes (↑).

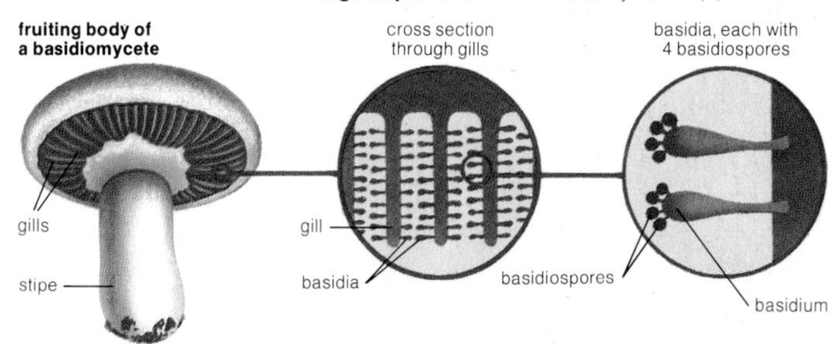

Basidiomycetes (*n*) a group of fungi (p. 163), recognized by their production of spores (p. 66) externally on a basidium (↓). Mushrooms (↓) and toadstools (↓) are the fruiting bodies (↓) of basidiomycetes.
basidium (*n*) the reproductive (p. 59) organ (p. 88) of a basidiomycete (↑). A basidium consists of one or four cells, bearing four basidiospores (↓) on short stalks. **basidia** (*pl.*).
basidiospore (*n*) the haploid (p. 50) spore (p. 66) of a basidiomycete (↑), produced on a basidium (↑).
fruiting body = fruit.
mushroom (*n*) the name for the reproductive (p. 59) structure in basidiomycete (↑) fungi (p. 163) of the family (p. 134) Agaricaceae.
toadstool (*n*) the fruiting body (↑) of a basidiomycete (↑) fungus (p. 163), consisting of a stalk and a cap. The cap has gills (p. 166) on the underside, on which the spores (p. 66) are produced. Toadstools are often poisonous.

toadstools and mushrooms

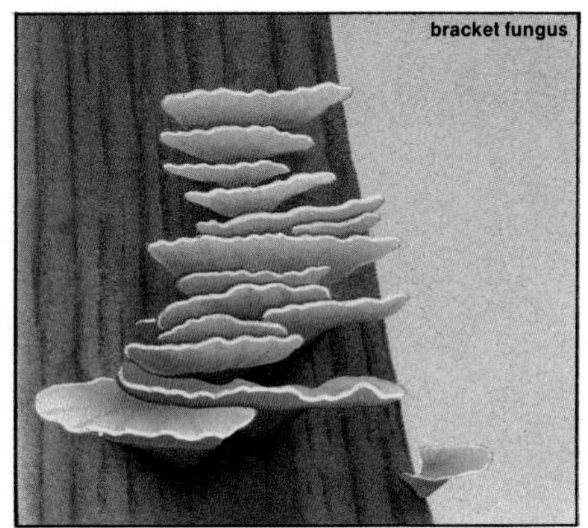

bracket fungus a basidiomycete (p. 165) fungus growing in the wood of living or dead trees, producing large flat-topped bracket-like fruiting bodies (p. 165) on the side of the host (p. 144).

stipe (*n*) a stalk, especially of a mushroom (p. 165) or toadstool (p. 165), or of a large seaweed (p. 122).

gill (*n*) a flat, vertically-positioned structure on the underside of the cap of a mushroom (p. 165) or toadstool (p. 165). The cap has many gills, radiating from the centre. The gills bear basidia (p. 165) on their surfaces.

dikaryon (*n*) a stage in the life cycle (p. 64) of many basidiomycetes (p. 165) when all the cells contain two haploid (p. 50) nuclei (p. 19). Each nucleus is derived from a different parent. **dikaryotic** (*adj*).

plasmogamy (*n*) fusion (p. 61) of the cytoplasm (p. 18) of two cells from different parents. This is the beginning of sexual (p. 59) reproduction (p. 59) in fungi (p. 163).

karyogamy (*n*) the fusion (p. 61) of two nuclei (p. 19) after plasmogamy (↑). In some fungi (p. 163), e.g. Basidiomycetes (p. 165), karyogamy takes place many cell divisions (p. 45) after plasmogamy. Between plasmogamy and karyogamy, each cell is a dikaryon (↑).

formation of a clamp connection in a basidiomycete

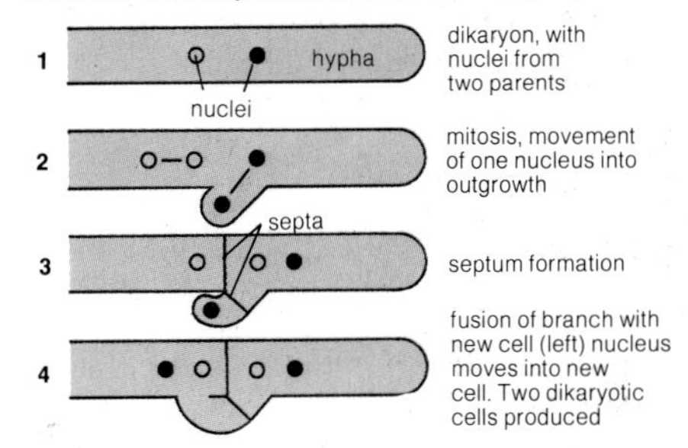

clamp connection a small looped branch of a hypha (p. 163) which grows at the time of cell division (p. 45) and septum (p. 163) formation in the dikaryon (↑) of a basidiomycete (p. 165).

dolipore septum a complex pore (p. 19) in the septum (p. 163) of a basidiomycete (p. 165) hypha (p. 163).

dolipore septum in a basidiomycete
(L.S. hypha)

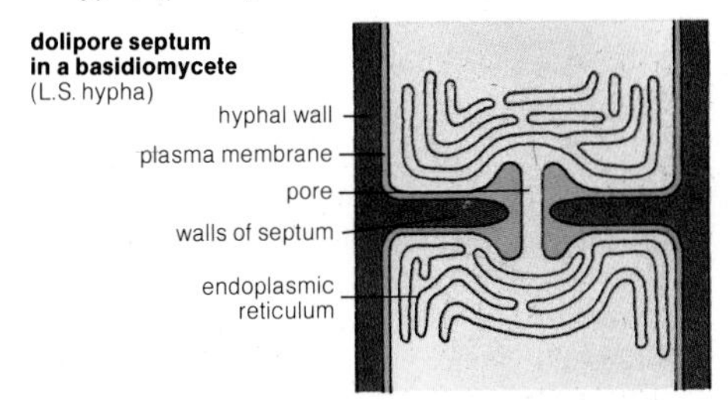

rust (*n*) a parasitic (p. 144) basidiomycete (p. 165) fungus (p. 163) of the order (p. 134) Uredinales. Rusts cause dark spots on the stems and leaves of plants. Some rusts, e.g. *Puccinia graminis* (cereal rust) are economically serious parasites.

uredospore (*n*) a type of vegetative (p. 60) spore (p. 66) produced by rust (↑) fungi (p. 163). Uredospores are dikaryotic (↑).

teleutospore (*n*) a type of thick-walled resting spore (p. 66) produced by rust (↑) fungi (p. 163). The teleutospore is the basidium (p. 165), eventually producing basidiospores (p. 165).

Zygomycetes (*n*) a group of fungi (p. 163) which produce non-motile (p. 121) spores (p. 66) in sporangia (p. 66).

homothallic (*adj*) of zygomycete (↑) species (p. 134) which exist in a single physiological (p. 111) form. Zygospores (↓) can be produced as a result of sexual (p. 59) fusion (p. 61) between identical mycelia (p. 163) growing together.

heterothallic (*adj*) of zygomycete (↑) species (p. 134) which exist in two different forms that appear identical but have different physiologies (p. 111). Zygospores (↓) are only produced if the two forms are growing together.

zygospore formation

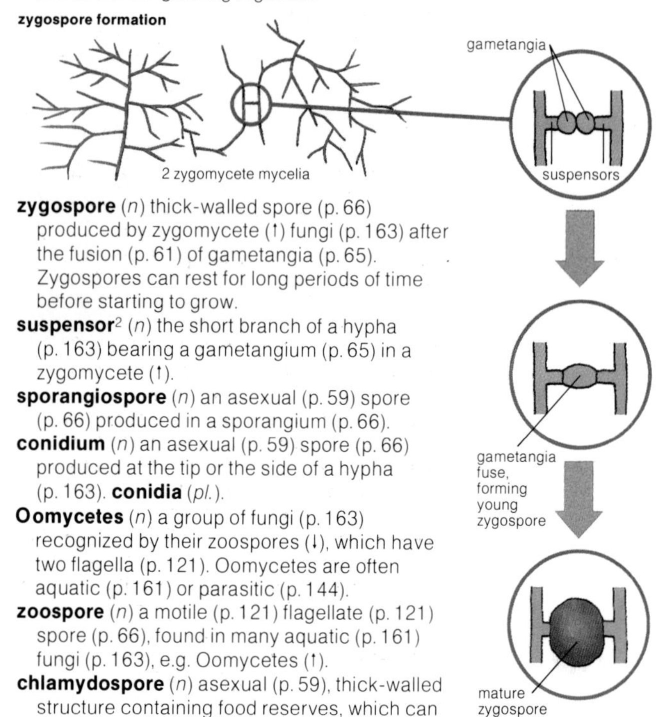

zygospore (*n*) thick-walled spore (p. 66) produced by zygomycete (↑) fungi (p. 163) after the fusion (p. 61) of gametangia (p. 65). Zygospores can rest for long periods of time before starting to grow.

suspensor[2] (*n*) the short branch of a hypha (p. 163) bearing a gametangium (p. 65) in a zygomycete (↑).

sporangiospore (*n*) an asexual (p. 59) spore (p. 66) produced in a sporangium (p. 66).

conidium (*n*) an asexual (p. 59) spore (p. 66) produced at the tip or the side of a hypha (p. 163). **conidia** (*pl.*).

Oomycetes (*n*) a group of fungi (p. 163) recognized by their zoospores (↓), which have two flagella (p. 121). Oomycetes are often aquatic (p. 161) or parasitic (p. 144).

zoospore (*n*) a motile (p. 121) flagellate (p. 121) spore (p. 66), found in many aquatic (p. 161) fungi (p. 163), e.g. Oomycetes (↑).

chlamydospore (*n*) asexual (p. 59), thick-walled structure containing food reserves, which can survive periods when hyphae (p. 163) cannot grow.

Chytridiomycetes (*n*) a group of aquatic (p. 161) and soil-dwelling fungi (p. 163), commonly unicellular (p. 119), which produce zoospores (↑).

oogonium (*n*) reproductive (p. 59) organ (p. 88) in some fungi (p. 163) and algae (p. 119), which produces female gametes (p. 61), or oospheres (↓). Oogonia (*pl.*) are multinucleate (p. 163).

oosphere (*n*) the female gamete (p. 61) produced in an oogonium (↑).

oospore (*n*) a dormant (p. 117), thick-walled zygote (p. 61), formed by the fertilization (p. 62) of an oosphere (↑).

columella[2] (*n*) the central part of a sporangium (p. 66) in some fungi (p. 163), e.g. the order (p. 134) Mucorales.

Myxomycetes (*n*) the taxonomic (p. 133) group containing the true slime moulds (↓), also known as the acellular (↓) slime moulds.

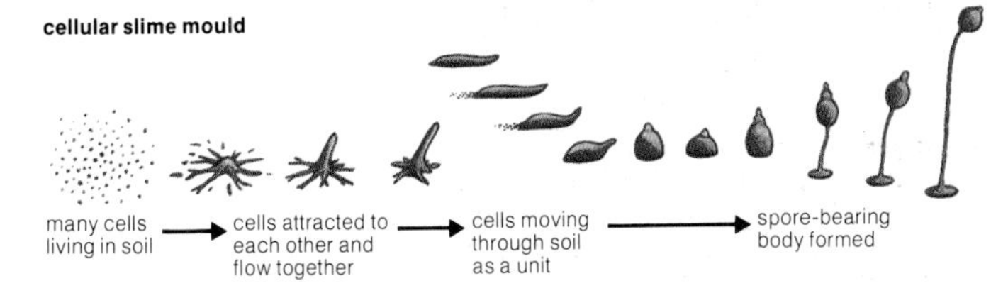

slime moulds a group of heterotrophic (p. 32) soil-dwelling organisms. They can be either acellular (↓) or cellular (↓). In acellular slime moulds, or Myxomycetes (↑), the organism is a plasmodium (↓). In cellular slime moulds, the cells are amoeba-like and single when feeding; when starved, they flow together to form a single sporulating (p. 66) structure.

plasmodium (*n*) a jelly-like multinucleate (p. 163) mass of protoplasm (p. 16) surrounded by a membrane (p. 18). This is the vegetative (p. 60) stage of an acellular (↓) slime mould (↑). A plasmodium can move through the soil.

acellular (*adj*) not made of cells, like the multinucleate (p. 163) plasmodium (↑) of a myxomycete (↑) slime mould (↑).

cellular (*adj*) made of cells.

structure (*n*) (1) the three-dimensional arrangement of the parts of a substance or organism, e.g. the structure of a molecule (p. 9) is the arrangement of its atoms (p. 8), the structure of a plant is the arrangement of its tissues (p. 88) and organs (p. 88); (2) any object with a definite shape or arrangement of parts, e.g. a molecule, a cell, a trunk (p. 92) of a tree.
function (*n*) the part played by a structure or a system, e.g. the function of chloroplasts (p. 36) is photosynthesis (p. 32), the function of photosynthesis (p. 32) is to produce carbohydrates (p. 28). **function** (*v*), **functional** (*adj*).
unit (*n*) (1) a single structure or object, which repeated many times forms a whole functioning object, e.g. the units of a nucleic acid (p. 51) are nucleotides (p. 52), the units of a population (p. 135) are individuals (p. 135); (2) a standard measurement, e.g. a metre, a kilogram.
sequence (*n*) (1) the order in which units are arranged one after another in a line, e.g. nucleotides (p. 52) in a nucleic acid (p. 51), amino acids (p. 56) in a protein (p. 56); (2) the order in which a set of chemical reactions (p. 11) in a metabolic pathway (p. 14) take place.
specialized (*adj*) of organisms and structures which are adapted for living in a particular habitat (p. 149) or shaped for a particular function, e.g. epiphytes (p. 137) are specialized for living on the branches of trees, leaves are specialized for photosynthesis (p. 32). **specialize** (*v*), **specialization** (*n*).
stimulus (*n*) an effect of the environment (p. 149) that causes, activates or quickens a process in an organism. A stimulus can be continuous, e.g. gravity, which causes the downward growth of roots, or periodic, e.g. light, which activates photosynthesis (p. 32), or sudden and occasional, e.g. damage, which activates the growth of callus (p. 108) tissue (p. 88) in plants.
modification (*n*) a minor change in structure or function, e.g. a bulb (p. 60) is an evolutionary (p. 139) modification of stem and leaves. **modify** (*v*).

mechanism (*n*) the way in which a process takes place, e.g. the mechanism of a chemical reaction (p. 11), or the way in which a functional unit works, e.g. the mechanism of an enzyme (p. 15).

medium (*n*) a solid or liquid substrate (p. 15), containing all the materials necessary for growth, used by biologists for the cultivation of organisms such as bacteria (p. 119), fungi (p. 163) and algae (p. 119), and also for the growth of tissue cultures (p. 69).

light microscope an instrument that uses light rays passing through a system of lenses to magnify small objects. The light microscope can be used for observing the arrangement of cells and tissues (p. 88) and the larger structures inside cells. It is not powerful enough for the observation of small details of cell structure. **microscopy** (*n*).

electron microscope a powerful instrument that uses electrons (p. 8) instead of light rays to magnify very small objects. The electron microscope can magnify more than 100 000 times, and can be used for observing very small details of cell structure.

stain (*n*) any one of many dye substances used in microscopy (↑) to show up particular parts of cells or tissues (p. 88).

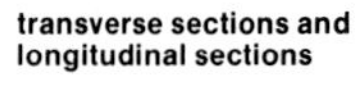

transverse sections and longitudinal sections

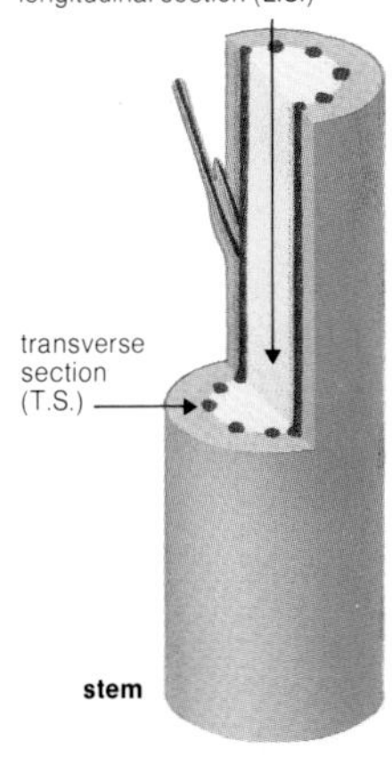

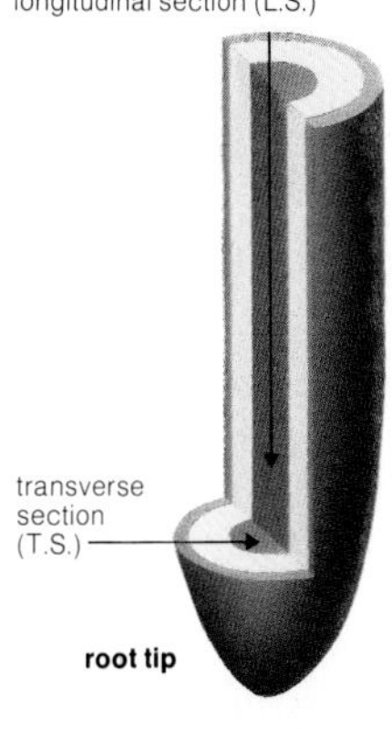

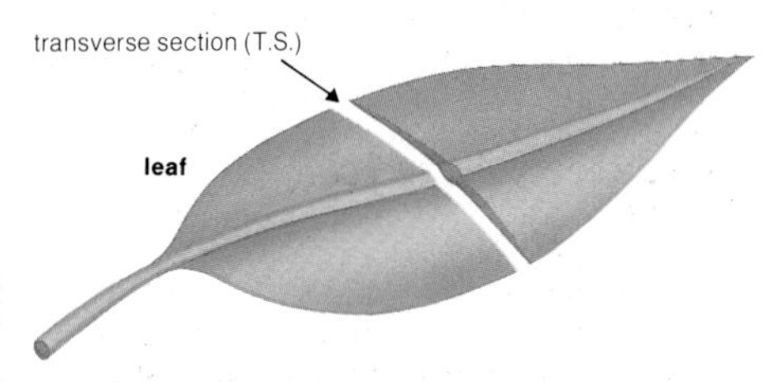

transverse section a cut made across an organ (p. 88) or tissue (p. 88), at right angles to the main direction of growth. T.S. (*abbr.*).

longitudinal section a cut made through or along an organ (p. 88) or tissue (p. 88), in the same direction as the main direction of growth. L.S. (*abbr.*).

mode (*n*) the most frequent (↓) value or class in a set of measurements or sample (↓).

mean (*n*) the arithmetic average of a set of measurements, defined by the equation

$$\bar{x} = \frac{x_1 + x_2 + x_3 + \ldots + x_n}{n},$$

where $\bar{x}$ is the mean, $x_1, x_2, x_3, \ldots, x_n$ are the individual measurements, and n is the number of measurements.

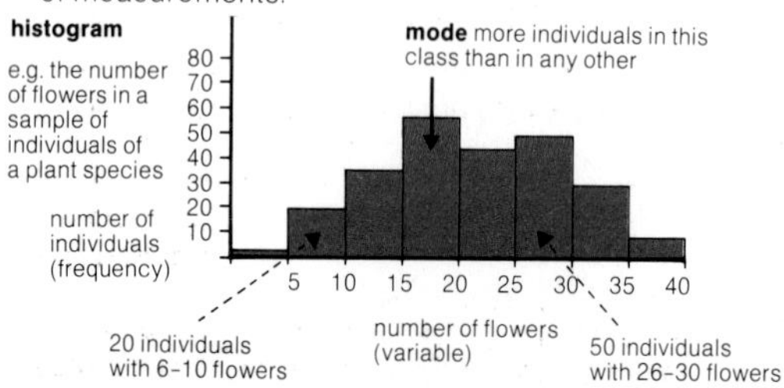

histogram (*n*) a way of showing the frequency (↓) with which different values of a variable (↓) occur in a sample (↓). The variable is divided into classes, and the frequency of each class is represented by the height of the bars on the chart.

normal distribution a symmetrical (p. 71) curve, showing the frequency (↓) of different values of a variable (↓) in a whole population (p. 135), i.e. the largest possible sample (↓). The mean (↑) and the mode (↑) are equal to each other in a normal distribution. Many biological variables are normally distributed.

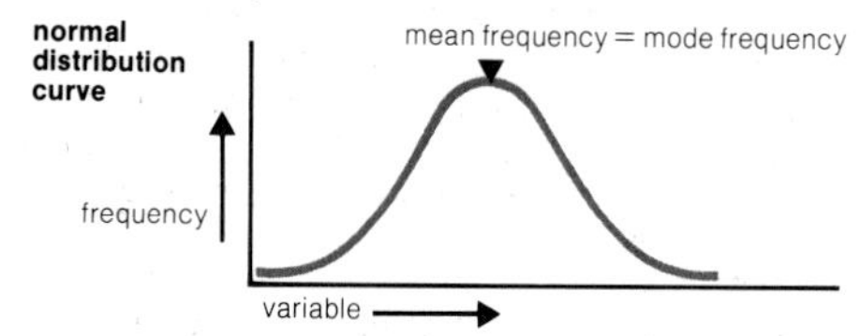

frequency (*n*) (1) a measure of how often an event takes place; (2) the number of times a particular class or value of a variable (↓) is recorded or observed in a sample (↓).

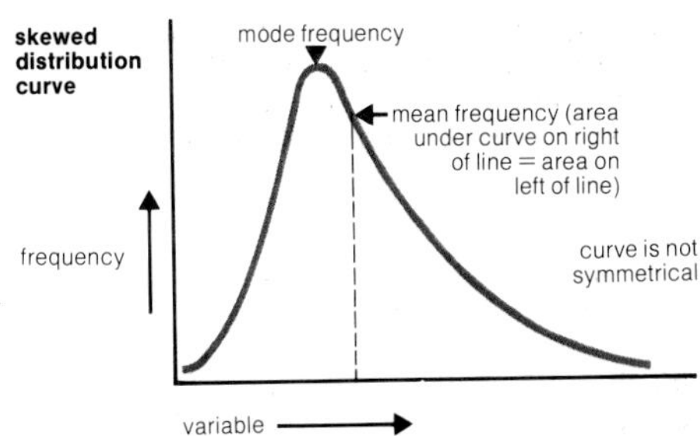

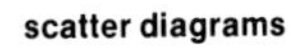

positive correlation
variable 2 increases as variable 1 increases

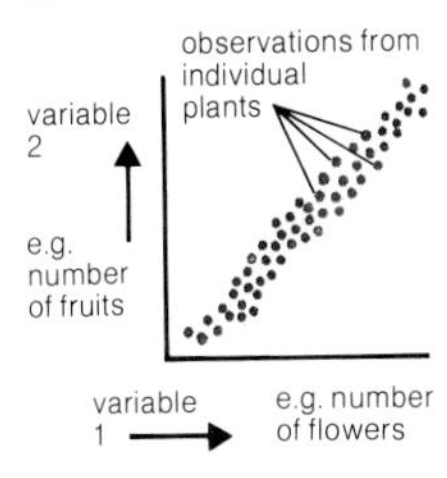

negative correlation
variable 2 decreases as variable 1 increases

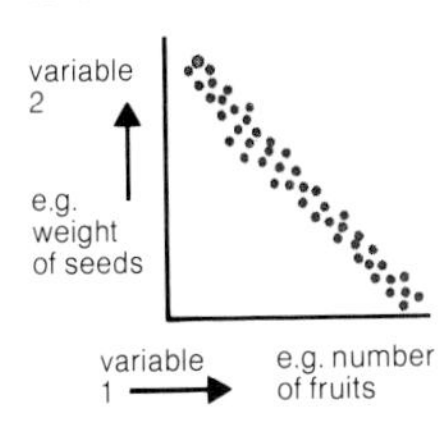

no correlation
variable 2 shows no relationship to variable 1

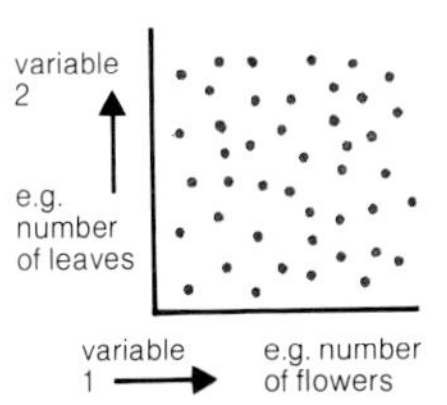

skewed distribution an asymmetrical (p. 71) curve, showing the frequency (↑) of different values of a variable (↓) in a whole population (p. 135). A skewed distribution differs from a normal distribution (↑) in that the mean (↑) and mode (↑) are unequal.

variable (*n*) a property or quantity that shows measurable variation, e.g. the weight of seeds in a fruit, the length of leaves on a shoot, the number of flowers on a plant.

scatter diagram a way of showing the relationship between two variables (↑), e.g. the relationship between the number of flowers produced by individuals (p. 135) of a plant species (p. 134) and the number of fruits they produce. Each point on a scatter diagram represents a pair of observations, one of each variable.

correlation (*n*) the process of determining whether the variation (p. 135) of one variable (↑) is related to the variation of another. If one variable increases at the same time as the other, they are said to be positively correlated. If one decreases as the other increases they are said to be negatively correlated. The correlation between two variables can be shown on a scatter diagram (↑).

sample (*n*) a small piece or part of a larger structure, area, or group, used by a scientist to measure or describe the properties of the larger object, e.g. quadrats (p. 162) are samples of vegetation (p. 150); dried plants in a herbarium (p. 133) are samples of species (p. 134).

APPENDIX ONE

Understanding botanical words

Many botanical and other scientific words or parts of words are derived from the Latin and Greek languages. The following pages contain some of the commoner word parts in the form of **prefixes**, that is, word parts that are added to the front of another word or word part to alter or specify its meaning. Many of the word parts in the list are not only used as prefixes, but also in the middle or at the end of words (sometimes in a slightly altered form); usually they cannot be used on their own. For instance, the prefix 'phyto-' (meaning: concerning plants) also appears as the word ending '-phyte' (meaning: plant), but not as a separate word.

Prefixes describing numbers or quantities are taken from Greek or Latin words. The following table shows the common prefixes from these two languages.

	GREEK PREFIX	LATIN PREFIX	PREFIX	MEANING	
1	mono-	uni-	hemi-	half	Gr
2	di-	bi-	semi-	half	L
3	tri-	ter-	poly-	many	Gr
4	tetra-	quad-	multi-	many	L
5	penta-	quinq-	omni	all	L
6	hexa-	sex-	dupli-	twice	L
7	hepta-	sept-	tripli-	three times	Gr
8	octo-	oct-	hypo-	less, under	Gr
9	nona-	novem-	hyper-	more, over	Gr
10	deca-	deci-	sub-	under	L
100	hecta-	centi-	super-	over	L
1000	kilo-	milli-	iso-	same, equal, identical	Gr

PREFIXES

a- without, not, lacking, e.g. *a*sexual, not sexual; *a*symmetrical, without symmetry.

ab- from, away, e.g. *ab*axial, the side of a leaf facing away from the stem.

ad- to, towards, e.g. *ad*axial, the side of a leaf facing towards the stem.

allo- different, differing, other, e.g. *allo*polyploid, a polyploid resulting from the fusion of two different nuclei; *allo*patric, of species occurring in different regions.

an- same prefix as **a-**, used before words beginning with a vowel or the letter h, e.g. *an*aerobic, not aerobic.

andro- male, e.g. *andro*ecium, the male parts of a flower.

anti- against, opposite, e.g. *anti*biotic, a substance that acts against living organisms (especially bacteria); *anti*podal, the cells at the opposite end of the embryo sac from the micropyle.

apo- from, away from, without, e.g. *apo*gamy, reproduction without sexual fusion; *apo*carpous, of flowers with carpels separate from one another.

auto- caused by or originating in itself, e.g. *auto*polyploid, a polyploid resulting from an increase in the number of sets of chromosomes within a nucleus; *auto*troph, an organism producing its own food.

bi- two, twice, double, e.g. *bi*nomial, the Latin name of a species, consisting of two words; *bi*ennial, a plant with a two-year life cycle.

bio- life, living, e.g. *bio*logy, the study of living things.

caul(i)- relating to stems, e.g. *cauli*florous, having flowers growing directly from the stem.

chromo- colour, coloured, e.g. *chromo*plast, a plastid containing pigments; *chromo*somes, so called because they become deeply coloured when stained for microscopy.

cleisto- closed, without an opening, e.g. *cleisto*gamy, self-pollination before the flower opens.

co- together, with, associated, e.g. *co*enzyme, a substance (not a substrate) that is necessary for the functioning of an enzyme.

crypto- hidden, e.g. *cryptc*phyte, a plant whose perennating organs are underground; *crypto*gam, a plant whose reproductive organs are very small or hidden.

cyto- relating to cells, e.g. *cyto*logy, the study of cells; *cyto*plasm, the parts of the cell outside the nucleus.

di- two, twice, double, e.g. *di*saccharide, a carbohydrate consisting of two sugar molecules (monosaccharides); *di*cotyledon, a plant with two cotyledons in the seed.

ecto- outside, e.g. *ecto*trophic mycorrhizae grow outside the cells of the host root.

endo- inside, inner, e.g. *endo*carp, the inner layer of the fruit wall; *endo*trophic mycorrhiza, with hyphae growing into the cells of the host root.

epi- on, upon, above, outer, e.g. *epi*carp, the outer layer of the fruit wall; *epi*phyte, a plant growing on another plant; *epi*geal germination, when the cotyledons emerge above the ground.

eu- good, normal, e.g. *eu*trophic, a habitat well-supplied or rich in nutrients.

ex- without, e.g. *ex*albuminous, without endosperm; *ex*stipulate, without stipules.

extra- outside, beyond, separate from, e.g. *extra*floral, away from the flower.

flavo- yellow, e.g. *flavo*protein, one of a group of yellow-coloured proteins.

gam(o)- joining together, fusion, e.g. *gamo*petalous, having fused petals.

gymno- naked, exposed, e.g. *gymno*sperm, a plant in which the seed is not enclosed in an ovary.

gyno- female, e.g. *gyno*ecium, the female parts of a flower.

halo- salt, salty, e.g. *halo*phyte, a plant that grows in salty habitats.

hemi- half, partly, e.g. *hemi*parasite, a parasite that produces some of its own food.

hetero- different, other, e.g. *hetero*zygous, with different alleles at the same locus on homologous chromosomes; *hetero*troph, an organism that obtains food other than from itself.

homo- same, similar, e.g. *homo*logous chromosomes have the same sequence of loci; *homo*sporous plants produce spores all of the same size.

hydro- relating to water, e.g. *hydro*phyte, a plant with perennating organs under water; *hydro*lysis, a chemical reaction involving the addition of water molecules and the breakdown of organic molecules.

hyper- more, above, very, e.g. *hyper*tonic, a more concentrated solution.

hypo- less, below, under, e.g. *hypo*tonic, a less concentrated solution; *hypo*gynous, a flower in which the corolla, calyx and anthers arise below the gynoecium.

infra- below, under, e.g. *infra*specific, variation below the level of species.

inter- between, e.g. *inter*specific, competition between species.

intra- within, e.g. *intra*specific, competition within a species.

iso- identical, e.g. *iso*gamy, the fusion of morphologically identical gametes.

lepto- thin, slender, e.g. *lepto*tene, the stage in the first meiotic prophase when the chromosomes appear as thin threads.

macro- large, great, long, e.g. *macro*molecule, a large molecule composed of many smaller molecular units.

mega- (1) great, large, e.g. *mega*spore, the larger of the two kinds of spore produced by heterosporous plants; (2) one million times.

meso- middle, between, e.g. *meso*phyll, the layer of tissue between the palisade and lower epidermis in a leaf; *meso*carp, the middle layer of the pericarp of a fruit.

micro- small, very small, e.g. *micro*scope, an instrument used for observing very small objects; *micro*spore, the smaller of the two kinds of spore produced by heterosporous plants.

mono- one, once, single, e.g. *mono*cotyledon, a plant with one cotyledon in the seed; *mono*carpic, a plant that produces fruit once in its lifetime.

morph(o)- shape, relating to shape, e.g. *morpho*logy, the study of shape.

multi- many, e.g. *multi*nucleate, of cells with many nuclei.

myco- relating to fungi, e.g. *myco*logy, the study of fungi.

neo- new, e.g. *neo*Darwinism, the science of evolution developed after Darwin, including the more recently-discovered principles of genetics.

oligo- few, e.g. *oligo*trophic, a habitat with few nutrients or low fertility; *oligo*saccharide, a carbohydrate consisting of a few monosaccharide units.

ortho- upright, correct, e.g. *ortho*tropic, an upright axis.

pachy- thick, fat, e.g. *pachy*tene, the stage in the first meiotic prophase when the chromosomes become short and thick.

palaeo- old, ancient, e.g. *palaeo*botany, the study of fossil plants.

pent(a)- five, e.g. *pent*ose, a monosaccharide with five carbon atoms.

peri- around, on the surface, e.g. *peri*anth, the parts of the flower around the reproductive parts; *peri*carp, the wall of the fruit.

photo- relating to light, e.g. *photo*synthesis, the production of carbohydrates using energy from light; *photo*tropism, curved growth towards light.

phyco- concerning algae, e.g. *phyco*biont, the algal partner in a lichen symbiosis.

phyll(o)- relating to leaves, e.g. *phyllo*taxy, the way in which leaves are arranged.

phyto- concerning plants, e.g. *phyto*chemistry, the chemistry of plants.

poly- many, e.g. *poly*peptide, a molecule with many peptide bonds.

rhiz(o)- relating to roots, rootlike organs or other underground plant parts, e.g. *rhiz*oid, the 'roots' of bryophytes; *rhiz*ome, an underground stem.

sapro- concerning decay, e.g. *sapro*phyte, a plant that lives on decaying organic matter.

schiz(o)- splitting, dividing, e.g. *schizo*carp, a fruit that splits into the separate carpels when ripe.

schler(o)- hard, rigid, e.g. *scler*enchyma, a hard, supporting plant tissue.

semi- half, partly, e.g. *semi*permeable, of membranes that allow the passage of some molecules but not others.

sub- under, below, somewhat, e.g. *sub*species, a taxon below the level of species; *sub*acute, a leaf apex that is somewhat acute.

sym- together, united, e.g. *sym*biosis, two different organisms living with and depending on each other.

syn- together, united, e.g. *syn*carpous, of ovaries in which the carpels are united.

tetra- four, e.g. *tetra*ploid, having four sets of homologous chromosomes.

tri- three, e.g. *tri*ose, a monosaccharide with three carbon atoms; *tri*ploid, having three sets of homologous chromosomes.

uni- one, once, single, e.g. *uni*cellular, an organism consisting of one cell.

xero- dry, e.g. *xero*phyte, a plant that occurs in dry habitats.

International System of Units (SI)

PREFIXES

PREFIX	FACTOR	SIGN	PREFIX	FACTOR	SIGN
milli-	$\times 10^{-3}$	m	kilo-	$\times 10^{3}$	k
micro-	$\times 10^{-6}$	μ	mega-	$\times 10^{6}$	M
nano-	$\times 10^{-9}$	n	giga-	$\times 10^{9}$	G
pico-	$\times 10^{-12}$	p	tera-	$\times 10^{12}$	T

BASIC UNITS

UNIT	SYMBOL	MEASUREMENT
metre	m	length
kilogramme	kg	mass
second	s	time
ampere	A	electric current
kelvin	K	temperature
mole	mol	amount of substance

DERIVED UNITS

UNIT	SYMBOL	MEASUREMENT
newton	N	force
joule	J	energy, work
hertz	Hz	frequency
pascal	Pa	pressure
coulomb	C	quantity of electric charge
volt	V	electrical potential
ohm	Ω	electrical resistance

SOME MULTIPLES OF SI UNITS HAVING SPECIAL NAMES

UNIT	SYMBOL	DEFINITION	MEASUREMENT
angstrom	Å	10^{-10} m = 10^{-1} nm	length
micron	μm	10^{-6} m	length
litre	l	10^{-3} m^3 = dm^3	volume
tonne	t	10^3 kg	mass
dyne	dyn	10^{-5} N	force
bar	bar	10^5 Pa	pressure

SOME NON-SI UNITS

UNIT	SYMBOL	DEFINITION	MEASUREMENT
atm	atm	101325 Pa, 1.01325 bar	pressure
degree Celsius	°C	K ($t_C = t_K - 273$)	temperature
million years	Ma, m.y.	10^6 years	time
billion (US) years	Ga	10^9 years	time

Index